CAMBRIDGE LIBRARY COLLECTION

Books of enduring scholarly value

Earth Sciences

In the nineteenth century, geology emerged as a distinct academic discipline. It pointed the way towards the theory of evolution, as scientists including Gideon Mantell, Adam Sedgwick, Charles Lyell and Roderick Murchison began to use the evidence of minerals, rock formations and fossils to demonstrate that the earth was older by millions of years than the conventional, Bible-based wisdom had supposed. They argued convincingly that the climate, flora and fauna of the distant past could be deduced from geological evidence. Volcanic activity, the formation of mountains, and the action of glaciers and rivers, tides and ocean currents also became better understood. This series includes landmark publications by pioneers of the modern earth sciences, who advanced the scientific understanding of our planet and the processes by which it is constantly re-shaped.

Meteorological Observations and Essays

Famed for his seminal work in the development of atomic theory, John Dalton (1766–1844) was a chemist and natural philosopher who served for years as professor of mathematics and natural philosophy at the New College, Manchester. Dalton was born into a Quaker family in the Lake District; his early interest in weather was inspired by a local instrument-maker and meteorologist. He began keeping a meteorological diary in 1787, and this 1793 book is one of his earliest publications. It contains not only meteorological observations but also speculations about their causes. Beginning with a description of the instruments needed to undertake such investigations, Dalton considers a variety of natural phenomena, finishing by offering various theories on the causes of the Aurora Borealis. This book also contains many of the ideas that would go on to be developed in his future research and publications, for which he is better known.

Cambridge University Press has long been a pioneer in the reissuing of out-of-print titles from its own backlist, producing digital reprints of books that are still sought after by scholars and students but could not be reprinted economically using traditional technology. The Cambridge Library Collection extends this activity to a wider range of books which are still of importance to researchers and professionals, either for the source material they contain, or as landmarks in the history of their academic discipline.

Drawing from the world-renowned collections in the Cambridge University Library, and guided by the advice of experts in each subject area, Cambridge University Press is using state-of-the-art scanning machines in its own Printing House to capture the content of each book selected for inclusion. The files are processed to give a consistently clear, crisp image, and the books finished to the high quality standard for which the Press is recognised around the world. The latest print-on-demand technology ensures that the books will remain available indefinitely, and that orders for single or multiple copies can quickly be supplied.

The Cambridge Library Collection will bring back to life books of enduring scholarly value (including out-of-copyright works originally issued by other publishers) across a wide range of disciplines in the humanities and social sciences and in science and technology.

Meteorological Observations and Essays

John Dalton

CAMBRIDGE UNIVERSITY PRESS

Cambridge, New York, Melbourne, Madrid, Cape Town,
Singapore, São Paolo, Delhi, Tokyo, Mexico City

Published in the United States of America by Cambridge University Press, New York

www.cambridge.org
Information on this title: www.cambridge.org/9781108184489

© in this compilation Cambridge University Press 2011

This edition first published 1793
This digitally printed version 2011

ISBN 978-1-108-18448-9 Paperback

METEOROLOGICAL

OBSERVATIONS

AND

ESSAYS.

BY

JOHN DALTON,

PROFESSOR OF MATHEMATICS AND NATURAL PHILOSOPHY,
AT THE NEW COLLEGE, MANCHESTER.

———————

EST QUODDAM PRODIRE TENUS, SI NON DATUR ULTRA.

HORACE.

———————

LONDON:

PRINTED FOR W. RICHARDSON, UNDER THE ROYAL EXCHANGE;
J. PHILLIPS, GEORGE-YARD; AND W. PENNINGTON, KENDAL.

1793.

———————

PRICE FOUR SHILLINGS.

PREFACE.

WHEN I first adopted the resolution to offer the public, in this manner, the result of my meteorological observations, which was about twelve months ago, my principal design was, to explain the nature of the different instruments used in meteorology, particularly the barometer and thermometer. As the number of these is increasing daily, many of them must fall into hands that are much unacquainted with their principles, and may therefore not profit by them in so great a degree as otherwise; for which reason, a short and clear explanation, with a series of observations serving further to illustrate and exemplify the principles, and a few practical rules for judging of the weather, deduced from experience, seemed to me to promise utility; whilst the observations themselves would be an addition to the stock already before the public, and might perhaps be found subservient to the improvement of the science.

Soon after this, having discovered the relation of the *aurora borealis* to magnetism, in the manner described in the introduction to that essay,

I found.

I found, that in order to eftablifh the difcovery,
a pretty large differtation would be required,
which muft, of courfe, be addreffed more peculiarly to philofophers; this neceffarily enlarged
the work, and became a primary confideration,
though the original defign was ftill kept in view;
I concluded afterwards, that the work fhould
confift of two parts, the firft of which was to
contain the fubftance of the original defign,
namely, a brief explanation of the nature of the
inftruments, and a digeft of all the obfervations
I had made, as matters of fact; the fecond was
to contain the effay or differtation on the *aurora*
borealis, together with fhort theoretic remarks
on the different phenomena of meteorology,
which I intended to felect chiefly from the beft
accounts I could procure; however, not having
by me all the books I could have defired, I was
neceffarily, and perhaps luckily, forced to con-
template a good deal on the different fubjects,
and to try fuch experiments as were within my
reach. The refult was, that feveral things occurred to me which were new, at leaft to myfelf,
and which throw light on the different branches
of natural philofophy, and of meteorology in
particular. Thefe I have thrown into the form
of Effays, in which are alfo given, fuch ufeful
difcoveries and obfervations of others as feemed
neceffary to be known, in order to form a proper
idea of the prefent ftate of the fcience, and of the
improvements that are yet to be made in it.

In

In the firſt part I have given not only the obſervations made at *Kendal* by myſelf, but alſo, with his leave, thoſe made at *Keſwick* by **Mr. *Croſthwaite*,** keeper of the muſeum at that place, together with obſervations on the barometer and rain, made at *London*, for three years, taken from the Philoſophical Tranſactions. The reſults of the ſeveral obſervations I have arranged and digeſted to the beſt of my judgment. The obſervations on the height of the clouds, and on the *aurora borealis*, particulaily the ſupplemental ones, are new, and, I ſuppoſe, in ſome reſpects, original, having never ſeen any other of a ſimilar nature publiſhed.

In the ſecond part, the firſt eſſay, though it contains little or nothing new, will be found a proper introduction to the ſubſequent ones.

The ſecond eſſay, containing the theory of the trade-winds, was, as I conceived when it was printed off, original; but I find ſince, that they are explained on the very ſame principles, and in the ſame manner, in the Philoſophical Tranfor 1735, by *George Hadley*, Eſq. F. R. S.—See Martyn's Abridgment, Vol. 8, part 2, page 500.

The third eſſay, on the variation of the barometer, I ſhould ſuppoſe will be conſidered as having ſome merit; it is new to myſelf, but as I am not well read in the modern productions on

the

the atmofphere, I cannot fay it will be found entirely fo to others. It may be proper to obferve,
that I had not adopted the theory of vapour
which is maintained in the fixth effay, when the
third was printed; but I know of no material
alteration I would have made in this effay, had
it been otherwife.

The fourth and fifth effays are chiefly felected
from the publications of others, except that in
the latter I have offered fome new thoughts on
the effect of the fituation of countries upon their
temperature.

In the fixth effay, amongft other things I have
advanced a theory of the ftate of vapour in the
atmofphere, which, as far as I can difcover, is
entirely new, and will be found, I believe, to
folve all the phenomena of vapour we are acquainted with; I have attempted to folve feveral,
particularly in the appendix.

In the feventh effay the relation betwixt the
barometer and rain is inveftigated, from the obfervations in the firft part; fome conclufions are
thence obtained in fupport of theory, and from
which feveral ufeful and practical obfervations
may be deduced.

The eighth effay is the large one on the
aurora borealis, which I have divided into fix
fections;

sections; this will no doubt attract the attention of philosophers. The reader will perceive all along, that I have spoken of the discovery therein contained as an original one; when I wrote the note at page 158, I had not seen the Abridgment of the Philosophical Transactions of the Royal Society; but I find from it that the learned and ingenious Dr. *Halley* formed an hypothesis to account for the *aurora borealis* by magnetism; in the Abridgment by *Jones*, Vol. 4, part 2, we find, that the Doctor, after enumerating particulars of several appearances, conjectures that they are occasioned by the earth's magnetism; and he endeavours to illustrate the hypothesis by placing a *terella*, or spherical magnet, with one of its poles upon an horizontal plane strewed with steel filings, which being done, the filings form various straight lined and curvilinear figures, according as they are situate near to or distant from the magnetic pole; these he thinks are analogous to the beams of the *aurora borealis.* The *light* of the *aurora* he is pretty much at a loss to account for, as electricity was then but imperfectly known.——If these hints of his had been pursued by others, the fact would undoubtedly before this have been established, *that the beams of the* aurora borealis *are governed by the earth's magnetism;* but instead of this, philosophers have amused themselves and others with forming various other theories to account for the phenomena, most of which are extravagant,

gant, not to fay ridiculous, M. *Mairan's zodiacal light* not excepted. Notwithftanding what the learned Doctor has fuggefted, I prefume it will be allowed, that the above mentioned fact has not hitherto been afcertained, unlefs it be done in the following work.

Whilft I am blaming others for framing fanciful theories, perhaps the cenfure may be retorted upon myfelf.—The fourth fection of the effay in queftion, entitled the ' theory of the *aurora borealis*,' will perhaps be regaided by many as wild and chimerical; but the *facts* which I have endeavoured to afcertain, refpecting the *aurora*, will excufe me for a momentary indulgence of the ideas of a vifionary theorift, if they be confidered as fuch.

The appendix contains the refult of barometrical and other obfervations to determine the height of *Kendal* and *Kefwick* above the fea, more exactly than is ftated in the preliminary remarks to the obfervations on the barometer; alfo, an account of the heights of fome mountains in the neighbourhood of *Kefwick*; it concludes with a further illuftration of the doctrine of vapour, and an explanation of fome facts relating thereto, particularly thofe obferved in working the air-pump.

It will be fufficiently evident that I have not had a fuperabundant affiftance from books, in
providing

providing and digefting the matter contained in
the following pages; by an attentive confideration of facts I have drawn conclufions in fome
inftances which had formerly been done, though
unknown to me at the time; thefe, however,
are fuch as would have been inferted had it
been otherwife, and therefore the defign of the
work is not in any manner fruftrated by the
circumftance*. At the fame time I acknowledge,
with particular fatisfaction, the friendly aid and
affiftance of one or two individuals, in the article
of books; to one perfon more particularly I am
peculiarly indebted, not only in this refpect, but
in many others; indeed, if there be any thing
new, and of importance to fcience, contained in
this work, it is owing, in great part, to my having had the advantage of his inftruction and
example in philofophical inveftigation.

I CANNOT help obferving here, that the following fact appears to be one of the moft remark-
b   able

---

* Since writing the above, I have met with an account of
Mr. *De Luc*'s elaborate work on the modifications of the atmofphere, (vid. the Appendixes to the 49th and 50th vols.
of the Monthly Review) from which it appears he maintains
nearly the fame principles in explaining the variations of the
barometer as I have done; his idea of *vapour* too feems not
unlike mine.—It is a favourable circumftance to any theory,
when it is deduced from a confideration of facts by two
perfons independently of each other.

able that the hiftory of the progrefs of natural philofophy could furnifh.———Dr. *Halley* publifhed in the Philofophical Tranfactions, a theory of the trade-winds, which was quite inadequate, and immechanical, as will be fhewn, and yet the fame has been almoft univerfally adopted; at leaft I could name feveral modern productions of great repute in which it is found, and do not know of one that contains any other. The fame gentleman publifhed, through the fame channel, his thoughts on the caufe of the *aurora borealis*, as mentioned above, which muft then have appeared the moft rational of any that could be fuggefted, and yet I do not find that any body has afterwards noticed it, except *Amanuenfis* (fee page 159). On the other hand, G. *Hadley*, Efq. publifhed in a fubfequent volume of the faid Tranfactions, a rational and fatisfactory explanation of the trade-winds; but where elfe fhall we find it?

Manchefter, Sep. 21, 1793.

ERRATA.

IN Part firft, the fections after the eighth are numbered wrong; they fhould be corrected as follows:

Page 52, for Section *tenth,* read Section *ninth.*
Page 54, for Section *eleventh,* read Section *tenth.*
Page 61, for Section *twelfth,* read Section *eleventh.*

SUBSCRIBERS

SUBSCRIBERS' NAMES.

A

THE Right Honourable Lady Vifcountefs Andover
Mr. William Allen, London

B

Mr. John Banks, Lecturer in Philofophy
Mr. Robert Barclay, Clapham
Mr. Robert Barclay, Clapham Terrace
Mr. Robert Barnes, Cockermouth
Rev. Thomas Barnes, D. D. Manchefter
Mr. Jofeph Beefley, Worcefter
Mr. George Benfon, Kendal
Mr. George Birkbeck, Settle
Mr. Robert Boyes, Carlifle
Mr. Richard Bradley, Manchefter
Mr. George Braithwaite, Kendal, 2 copies
Mr. George Browne, Troutbeck

C

Mr. Colin Campbell, Kendal
Mr. Ecroyde Claxton, Amblefide
Dr. Clealby, Barnardcaftle
Mr. Peter Clare, Manchefter
Mr. Jofeph Cockfield, Upton
Mr. William Cockin, Burton
Mr. John Cole, Rochefter
Mr. William Coward, Kendal
Mr. William Bell Crafton, Tewkefbury
Mr. Samuel Crewdfon, Kendal
Mr. Simon Crofield, Kendal
Mr. Peter Crofthwaite, at the Mufeum, Kefwick, 6 copies

D

Mrs. Deborah Dalton, Kendal
Mr. Jonathan Dalton, Kendal, 2 copies
Mr. John Dalton, Eaglesfield
Mr. George Davis, Blackheath
Mr. John Dawson, Sedbergh
Mr. Robert Dickinson, Kendal
Mr. Jonathan Dixon, Lowes-water

F

Mr. John Fallowfield, Preston
Mr. John Farrer, Grayrigg
Mr. James Fell, Kendal
Mr. William Ferguson, Kendal
Mr. William Field, Jun. Cartmel
Mr. John Fletcher, Greysouthen

G

Mr. Samuel Gawthrop, Kendal
Mr. Robert Goad, Jun. Haws
Mr. John Gough, Kendal
Mr. Robert Greenhow, Kendal
Mr. Thomas Greenup, London
Mr. Joseph Gurney, Norwich

H

Mr. Ponsonby Harris, Eaglesfield
Mr. Benjamin Harrison, Jun. Kendal
Mr. George Harrison, London
Mr. Thomas Harrison, Kendal
Mr. Thomas, Hoyle, Jun. Manchester
Mr. John Hayes, Proctor, Durham
Mr. Isaac Hewetson, Jun. Penrith
Mr. John Hodgson, Bampton
Mr. David Hodgson, Wormanby, near Carlisle
Mr. Thomas Holme, Kendal
Mr. Anthony Horne, London
Honourable Mrs. Howard
Charles Hutton, LL. D. and F. R. S. of the Royal military
    Academy, Woolwich

I

I

Mr. John Ireland, Kendal

J

Mr. James Jackson, Ketley, Shropshire

K

Charles Kerr, Esq. Abbotrule
Mr. Justinian Kerry, Stainton
Mr. Thomas Kitching, Kendal

L

Mr. —— Lloyd, Lecturer in Philosophy

M

Mr. Thomas Holme Maude, Kendal

P

Mr. James Parratt
Mr. Samuel Parrat, Deer-park
Mr. Jonathan Peele, Cockermouth
Mr. Joseph Pennington
Thomas Percival, M. D. F. R. S. A. S. &c. Manchester
Mr. James Phillips, Bookseller, London, 25 copies
Mr. John Phillips, Whitehaven, 2 copies
Joseph Pocklington, Esq.
Rev. John Pope, Manchester
Mr. Joseph Priestley, Bradford

R

Rev. Jeremy Reed, Rookcliffe, near Carlisle
Mr. William Reynolds, Ketley bank, Shropshire, 7 copies
Mr. William Richardson, Kendal
Mr. —— Robinson, Bookseller, Wigton, 2 copies
Mr. Edward Robinson, Kendal
Mr. Elihu Robinson, Eaglesfield
Mr. Edward Rogers, Liverpool
Mr. William Rooke, Newton
Rev. Caleb Rotheram, Kendal

S

Mr. Richard Sleddall, Kendal
Mr. John Smellie, London
Mr. John Sowden, Kendal

The

The Right Honourable the Earl of Stanhope, F. R. S.
Jarrard Strickland, Efq. Kendal, 2 copies
Mrs. Strickland, Kendal, 2 copies
Mr. William Strickland, Kendal
Mr. John Swainfton, Kendal

T

Mr. John Thomfon, Kendal
Mr. John Townfhend, Jun. London
Mr. Benjamin Townfon

W

Mr. Edward Wakefield, Kendal
Mr. William Wakefield, Kendal
Mr. John Walker, Teacher of the Claffics and Mathematics,
    Ufher's Ifland, Dublin, 2 copies
Mr. Peter Walker, Bookfeller, Cockermouth
Mr. Jofeph Waugh, Jun. Cockermouth
Mr. Ifaac Whitwell, Kendal
Mr. John Wilkins, Cirencefter
Mr. Chriftopher Wilfon, Jun. Kendal
Mr. Robert Wilfon, High-Scales, near Orton
Mr. —— Wilfon, Burnefide
Mr. Peter Winder, Rogerfcale
Mr. John Wright, Efher.

CONTENTS.

(xv)

CONTENTS.

PART I.

METEO-

METEOROLOGICAL

OBSERVATIONS AND ESSAYS.

PART FIRST,

OBSERVATIONS.

SECTION FIRST,

Of the Barometer.

THE barometer, or common weather-glafs, confifts of a ftraight glafs tube, above 31 inches long, and open at one end, that has been filled with quickfilver, and afterwards inverted into a bafon of the fame fluid, by applying a finger to the open end, fo as to exclude all air from entering the tube; in this cafe, the finger being withdrawn, and the tube erected, the quickfilver leaves the top of it, and finks fo as to ftand at the height of about 29 or 30 inches

above

above the furface of that in the bafon; it is then
applied to a frame, with a fcale graduated fo as
to mark at all times the height of the column,
in inches and tenths, &c. The inftrument thus
completed is called a barometer.—It is ufual
now to blow a pretty capacious bulb at the open
end of the tube, and then bend the tube a little
above the bulb, fo that the bulb may ftand upright, leaving a little orifice in it to admit the
quickfilver; then the tube being filled as before,
upon being inverted, the column of quickfilver
in the tube ftands at the height of 29 or 30
inches above the furface of that in the bulb, as
in the former cafe.

The reafon of the fact may be explained thus:
every body that fupports another, bears all its
weight; therefore when the furface of any non-
elaftic fluid is expofed to the air, it bears the
weight of a column of air whofe bafe is equal to
the faid furface, and its height that of the atmofphere, fuppofed to be 40 or 50 miles; now
though air be a very light fubftance, being in its
ufual ftate, at the earth's furface, about $\frac{1}{800}$th
part of the weight of an equal quantity of water,
yet fo prodigious a column of it as that above
mentioned, has a very confiderable weight;
moreover, it is a fundamental principle in hydroftatics, that the preffure upon the furface of
a fluid muft be the fame on each part, or the
fluid will not reft till that is the cafe; if, therefore,

fore, the preffure be removed from any place of
the furface, either wholly or in part, the fluid
will yield in that place, and afcend till the weight
of the column of fluid above the furface, together with the preffure upon the column, if any,
are equal to the general preffure upon the fluid
in every other part.—In the cafe of the barometer, there is a *vacuum* at the top of the column,
and confequently no preffure upon its furface, fo
that the weight of the column alone balances the
preffure of the atmofphere without, upon the
furface of the fluid in the bafon. This *equilibrium*, between the mercurial column and column of air, is very clearly illuftrated and confirmed by means of the air-pump; for, when a
barometer is inclofed in a receiver, as the air is
exhaufted, and its preffure of confequence decreafed, the mercurial column defcends proportionally. It appears then, that the weight of the
air fupports the mercury in the barometer, and
that the weight of the mercurial column is equal
to the weight of a like column of air extending
to the top of the atmofphere.——When the
tube is bent at the bottom, and turned up, the
fame reafoning, joined to the principle that fluids
in bent tubes always rife to the fame height in
each leg, when they are both open to the atmofphere, will explain the fact in this cafe.

From thefe confiderations, the weight of the
whole atmofphere may be readily found; for, it

is equal to the weight of a quantity of quickfilver fufficient to cover the whole furface of the
globe to the height of 30 inches nearly.

The great weight and preffure above afcribed
to the atmofphere, and their effects, are affented
to by philofophers of the prefent age, without
fcruple; but people not much verfant in philofophical enquiries admit them with reluctance;
they apprehend, that if bodies were preffed with
the force above mentioned, which amounts to
about 15lbs. avoirdupoife upon each fquare inch
of furface, the effect fhould be obvious; whereas
it is found that bodies of the flighteft texture
are unhurt by the atmofphere,—and the great
facility with which bodies are moved in the atmofphere, they conceive as another objection.—
Perhaps it may be fome help to thefe to obferve,
that the atmofphere preffes equally upon bodies
in every direction, and has therefore no tendency
to feparate their parts; and, as for the refiftance
which bodies meet with in moving through the
atmofphere, it is not proportionate to the *preffure*
of the atmofphere, but to its *denfity*, which
being very little, as has been obferved above,
the refiftance is fmall.

The barometer was invented in 1643, by
Torricelli, at Florence, in Italy.—The phenomenon foon attracted the notice of philofophers
of that age, and the more fo, as it feemed to
prove

prove the exiftence of a *vacuum,* when the opinion of its non-exiftence was general, and the
maxim that *nature abhors a vacuum,* was almoft
unqueftioned. Had the quickfilver ftill continued to fill the tube when erected, the fact
would have been accounted for on this imagi-
nary principle, and have paffed without further
notice. As it was, however, thofe who ftill ad-
hered to the maxim were reduced to great difficulties, and forced to have recourfe to various
unmeaning fubtilties, to get rid of the *vacuum;*
whilft many began to queftion the truth of the
maxim itfelf. At length it was clearly proved,
from the inftance in queftion, and from other
phenomena, that the maxim was contradictory
to the laws of nature; the fufpenfion of the mercury in the barometer was attributed to its true
caufe, the weight of the air; and the fpace at
the top of the tube was afcertained to be nearly
a perfect *vacuum,* or fpace void of matter. This
difcovery, as it led to that of the air-pump, and
other important ones, is juftly regarded as one
of the greateft in the laft century.

Torricelli, the inventer of the barometer, obferved, that if it was fuffered to ftand for a length
of time, the height of the mercury in the tube
was perpetually varying, though its whole range
did not exceed 2 inches at that place; it was
further noticed, that this variation feemed to
have fome affinity to the weather, the quickfilver
being

being generally low in windy and rainy weather, and high in ferene and fettled weather, which circumftance procured the inftrument the name of *weather-glafs.* This difcovery promifed to be of the utmoft importance to mankind, by enabling them to forefee thofe changes in the atmofphere, the knowledge of which was fo interefting to them, and the moft fanguine expectations were entertained on the fubject. The experience of a century and a half has now been obtained, from which the barometer does not feem to be that infallible guide that it was once expected to be, though it is certainly a very ufeful inftrument in this refpect, in the hands of a judicious and fkilful obferver.—But of this more hereafter.

SEVERAL ingenious contrivances have been ufed, by different perfons, to make barometers of a more ample range, in order to obferve minute alterations of weight in the atmofphere; but all thefe are liable to fuch objections as render the common upright one preferable.

Thofe who wifh to make barometrical obfervations, in order to compare them with others, fhould be well affured of the accuracy of their inftruments;—fuch as incline to make their inftruments themfelves would do well to attend to the following particulars,

That

That the tube be not lefs than one-eighth of an inch diameter within.

That the quickfilver be ftrained through a cloth, or rather through leather.

That the infide of the tube be perfectly dry, and the quickfilver dry when put into it.

If there be any moifture in the tube or quick-filver, it expands into an elaftic vapour when the preffure of the air is removed, and, afcending into the *vacuum*, depreffes the mercurial column fometimes to the amount of one-quarter of an inch, or more, below its proper ftation. The criterion to difcover moifture is to apply the warm hand to the *vacuum*, and the mercury will fink confiderably; but it will not be affected if clear of moifture.—Alfo, if upon a gentle agi-tation of the barometer in the dark, there appear a light in the *vacuum*, it is a fign there is little or no moifture. If, upon a gentle inclination, the quickfilver rife to the top of the tube, and completely fill it, leaving no fpeck, it is clear of air.

The fcale in ftrictnefs ought not to be full inches, but fomething lefs, owing to the rifing and falling of the furface of the refervoir. If the tube have a bulb, then the area of the furface at the top of the column, divided by the fum of the

areas

areas of the top and refervoir, will give the part to be deducted; but if the tube be ftraight, then the whole area of the refervoir, leffened by the area of the glafs annulus, made by a horizontal fection of the erected tube, muft be ufed as the denominator of the fraction; hence, if the fraction be $\frac{1}{60}$, then the fcale of 3 inches muft be diminifhed by half a tenth.

PREVIOUS to the detail of obfervations, it will be proper to defcribe the fituation of the places of obfervation. ——The latitude of *London* is 51° 31′ N.—*Kendal* is fituate in lat. 54° 17′ N. long. 2° 46′ W. There lies an extenfive range of mountains from it in every direction, except to the fouth. Their height may be from 1 to 6 or 7 hundred yards*; fome are near, but from the north to the eaft their diftance is 3, 4, or 5 miles. *St. George's Channel* bears SW. and the high water at fpring comes up the river to within 6 miles of the town, but low water is at a great diftance. The town may be 25 yards above the level of the fea.—*Kefwick* is fituate in lat. 54° 33′ N. long. 3° 3′ W.; it is well known to be in the centre of a mountainous country, and fome of the higheft mountains in the north of England are in its neighbourhood. It is 16 miles from the *Channel,* and perhaps about 45 yards above its level †.

The

---

* *Benfon-knot* is 310 yards above the level of the river; *Whinfel-beacon* is 500 yards above the fame; and *Kendal-fell* from 1 to 2 hundred yards.

† It may not be amifs to remind the young Tyro here, that the higher any place is above the level of the fea, the lower will the mean ftate of the barometer be at that place.

Of the Barometer.

The observations at *Kendal* were made by the author, three times each day, namely, betwixt 6 and 8 o'clock in the morning, at noon, and at 8 or 10 in the evening.—Those at *Keswick* were likewise made three times each day, the morning and noon observations about the same time as at *Kendal*, but the evening observations were made at 4 or 5 in the winter, and 6 in the summer; the observer was Mr. *Crosthwaite*, a gentleman, who, besides his attention to meteorology, has been for several years past assiduously furnishing a *museum*, for the entertainment of the tourists, at present consisting of a great variety of natural and artificial productions from every quarter of the globe, fossils, plants, &c. and he has also made accurate surveys of the lakes.

The observations at *London* are taken from the Philosophical Transactions of the Royal Society, being those made there by order of the president and council; they are made twice a day, namely at 7 A. M. (in December, January, and February at 8 A. M.) and at 2 P. M.

With respect to the barometers at *Kendal* and *Keswick*, they were both clear of air and moisture, and exhibited the electric light in the dark. The scales were both *full* inches, and therefore the variations were somewhat greater than the observations denote them.—About $\frac{1}{30}$ should have been allowed upon them.

In the following account we have given the mean state of the barometers, at the respective places, for each month of the year, and likewise for the whole year, together with the highest and lowest observations each month, and the time they took place; as also the direction of the wind, and its strength, at the time: the direction we have usually referred to some one of 8 equidistant points of the compass, and the strength is denoted by the figures 0, 1, 2, 3, and 4, respectively, the first marking a calm, or very gentle breeze, and the last a hurricane.

C

The

The obfervations at *London* are only for 3 years, becaufe the later ones could not be procured; thofe at *Kendal* and *Kefwick* for 5 years, from 1788 to 1792, inclufive. To the end of thefe is added the mean monthly ftate of the barometer, found from the means of the 5 years, as alfo the mean upon the extremes, the former of which is corrected, on account of the variations of heat in the different months, by which the quickfilver in the barometer is contracted or dilated, though retaining the fame weight.—We have alfo fummed up the fpaces defcribed by the quickfilver each month, noted the number of changes from afcent to defcent, and the contrary, and found their amount for the year.

By the *mean ftate*, applied to obfervations, is to be underftood the fum of all the particular obfervations divided by their number.

The upper part of the following tables, having no abbreviations, is fufficiently explicit; and in the under part, which contains the days in the feveral months on which the higheft and loweft obfervations were taken, and the winds at thofe times, we have ufed H, for higheft, L, for loweft, m, for morning, n, for noon, and nt, for night.

N. B. *Kendal* bears N. 30° W. from *London*, diftant nearly 226 Englifh miles, meafured on a great circle of the earth; *Kefwick* bears N. 35° W. from *Kendal*, diftant 22 Englifh miles, meafured on a great circle.

1788.

	LONDON			KENDAL			KESWICK		
	Mean	highest	lowest	Mean	highest	lowest	Mean	highest	lowest
Jan.	29.97	30.70	28.89	29.87	30.56	28.38	29.82	30.56	28.35
Feb.	29.68	30.21	28.65	29.47	30.22	28.65	29.42	30.17	28.61
March	29.68	30.08	29.32	29.56	30.09	29.15	29.51	30.07	29.12
April	30.07	30.48	29.50	29.95	30.41	28.97	29.89	30.36	28.92
May	30.04	30.34	29.58	30.02	30.41	29.47	29.94	30.32	29.37
June	29.94	30.28	29.49	29.94	30.31	29.50	29.89	30.26	29.46
July	29.99	30.22	29.73	29.82	30.12	29.47	29.76	30.12	29.40
Aug.	29.95	30.45	29.22	29.83	30.37	29.19	29.77	30.37	29.14
Sep.	29.86	30.25	29.37	29.74	30.16	29.28	29.67	30.09	29.20
Oct.	30.32	30.55	29.64	30.07	30.62	29.50	30.02	30.63	29.43
Nov.	30.11	30.50	29.61	29.98	30.34	29.22	29.92	30.32	29.20
Dec.	29.92	30.33	29.50	29.92	30.28	29.53	29.90	30.23	29.50
Inches	29.96 annual mean			29.85 annual mean			29.79 annual mean		

Month		LONDON		KENDAL		KESWICK	
Jan.	H	16 n	WNW 1	16 n & nt	W 1	16 n & nt	W 2
	L	3 n	S 2	3 n	S 2	4, m	W 1
Feb.	H	7 m & n	NE 1 (a)	7 n	SE 0	7 n	calm
	L	21 n	SSE 1	21 n & nt	NE 2 (b)	21 nt	SE 1
Mar.	H	3 n	E 2 (c)	3 n & nt	NE 1	3 n & nt	W 0
	L	23 m	N 2	1 m	SE 1	1 m	SW 0
Apr.	H	9 m	WNW 1	9 n	calm	9 n	S 2
	L	3 n	W 2	3 n	NW 3	3 n	W 3
May	H	3 m	ENE 2	3 all day	NE 2	3 m & n	E 2
	L	29 m	SW 1	9 n	SW 3	9 n	SW 3
June	H	5 n	W 1	9 all day	NE 1	9 n & nt	NE 1
	L	27 n	S 1	26 nt & 27 m	NW 0	27 m	NW 0
July	H	21 n	NW 1 (d)	21 n	NW 0	21 n	W 1
	L	5 n	SSW 1	16 m	SW 1	16 m	SW 1
Aug.	H	2 m	N 1	2 m	NW 1 (e)	2 n	NW 1
	L	14 m	SW 1	14 m	W 0	14 m	W 1
Sep.	H	12 m	ESE 1	11 nt	NW 0	11 nt	NW 0
	L	21 m	calm	29 n	W 2	29 n	W 2
Oct.	H	8 m	SEbS 2 (f)	8 n & nt	NE 0	8 n & nt	E 0
	L	16 n	W 1	16 m & n	SW 0	16 m & n	W 1
Nov.	H	1 m	E 1	1 m	SW 0	1 m	SE 0 (g)
	L	4 m	SW 2	3 nt	SW 2	3 nt	SW 3
Dec.	H	30 m	E 1	28 nt	N 0	28 nt	NE 0
	L	14 n	NE 1	31 nt	calm	31 nt	calm

(a) And 12 m. SW 1. (b) And 22 m. NE 2. (c) And 11 n. ENE 1.
(d) And 31 m. SW 1. (e) And 2 n. SE. 1. (f) And 8 n. NE 2,
(g) And 16 m. NE 0.

1789.

	LONDON.			KENDAL.			KESWICK.		
	Mean	highest	lowest	Mean	highest	lowest	Mean	highest	lowest
Jan.	29.72	30.75	28.58	29.59	30.75	28.12	29.51	30.72	28.09
Feb.	29.70	30.34	28.65	29.52	30.19	28 50	29.44	30.20	28.43
March	29.72	30.13	28.94	29.71	30.12	28.87	29.59	30.13	28.71
April	29.77	30.18	29.10	29 64	30.09	28.94	39.51	29.98	28.77
May	29.88	30.27	29.57	29.77	30.19	29.37	29.66	30.12	29 23
June	29.84	30.23	29.40	29.77	30.20	29.25	29.66	30.12	29.14
July	29 85	30.09	29.54	29.74	30.00	29.50	29.63	29.88	29.40
Aug.	30 06	30 33	29.70	29 99	30.32	29 62	29.88	30.23	29.46
Sept.	29.88	30.38	29.30	29.75	30.25	29 25	29.64	30 15	29.17
Oct.	29.52	30.29	29.00	29.56	30.30	28.59	29 46	30.20	28.48
Nov.	29.70	30.46	28.72	29 60	30.34	28.69	29.48	30.27	28.60
Dec	29 86	30.56	28.88	29.63	30.41	28.72	29.48	30.32	28.57
Inches	29.79 annual mean			29.69 annual mean			29.58 annual mean		

		LONDON		KENDAL		KESWICK	
Jan.	H	5 n	ENE 1	5 nt	N o	5 n & nt	E o
	L	18 n	S 2	18 n	S 3	18 n	SW 2
Feb.	H	17 m	SW 1	16 nt	calm	16 nt	NW 1
	L	25 n	SW 2	25 n & nt	SW 1	25 n & nt	NE1 N2
Mar.	H	3 m	NNE 1	6 m & n	NE 1	6 m & n	NE1 N1
	L	13 n	W 1	13 m & n	NE2 E3	13 n	SE 1
Apr.	H	21 m	WSW 1	9 nt	SW o	9 all day	E o
	L	3 m	SW 1	26 n	SW 3(a)	26 m & n	W1 (b)
May	H	19 n	W 2	11 n & nt	SW o	11 nt	W 1
	L	31 m	SW 1	17 nt	SW 2	17 nt	S 2
June	H	13 m & n	ENE 1	13 n & nt	N o	13 n & nt	S 1
	L	4 m	WSW 2(c)	22 m	SW o	22 m	E o(d)
July	H	1 m	W 1	28 & 30 nt	calm	28 & 30 nt	W1 So
	L	13 m & n	ESE 1	17 n	W 1	17 n	S o
Aug.	H	18 m	NNW 1	17 m	NE o	17 all day	NE o
	L	21 n	W 1(e)	22 & 31 nt	calm	31 nt	calm
Sept.	H	12 n	WNW 1	12 n	SW 1	12 n·	NW 1
	L	19 n	WNW 1	19 n	N o	19 nt	NW o
Oct.	H	27 n	NNW 1	27 nt	NE 1	27 nt	N 1(f)
	L	6 m	W 2	1 n	SW 3	1 m	S 2
Nov.	H	27 m	W 1	27 all day	N o	27 all day	NW o
	L	7 m	N 1	6 n & nt	NE 2	6 nt	N 3
Dec.	H	9 m & n	W 1	9 all day	SW o	9 nt	W 1
	L	15 m	W 2	15 n	W 1	15 n	SW 2

(a) And 27 m. SE 1.   (b) And 27 m. SE o.   (c) And 4 n. also 22 n. S2.
(d) And 4 m & n. W 1.   (e) And 22 m. W 1. also 31 m & n. SSE 1.
(f) And 28 m. NW o.

1790.

	LONDON.			KENDAL.			KESWICK.		
	Mean	highest	lowest	Mean	highest	lowest	Mean	highest	lowest
Jan.	30.07	30.47	29.27	29.91	30.34	28.65	29.89	30.36	28.64
Feb.	30.25	30.62	29.88	30.06	30.41	29.47	30 02	30.41	29.40
March	30.26	30.65	29.83	30.18	30.59	29.57	30.15	30.59	29.48
April	29.86	30.30	29.38	29.85	30.28	29.28	29.81	30 28	29.19
May	29.90	30.14	29.50	29.85	30.28	29.25	29.82	30.28	29.19
June	30.03	30.35	29.49	29.89	30.25	29.31	29.87	30.28	29.22
July	29.84	30.20	29.29	29.72	30.09	29.34	29.68	30.15	29.28
Aug.	29.97	30 16	29.64	29.81	30.03	29.47	29.77	30.05	29.42
Sept.	30.00	33 42	29.31	29.87	30.34	29.25	29.84	30 34	29.14
Oct.	29.89	30.40	29.62	29.80	30.28	29.22	29.75	33.26	29.11
Nov.	29.81	30.40	29.02	29 75	30.34	28.97	29.70	30.28	28.90
Dec.	29.88	30.38	28.80	29.70	30.34	28.84	29.67	30.31	28.74
Inches	29.98 annual mean			29.87 annual mean			29.83 annual mean		

Month		LONDON	KENDAL	KESWICK
Jan.	H	7 n WNW 1	7 n & nt calm	7 n&nt E 1, S 0
	L	28 n SW 2	28 nt SW 2	28 nt W 0
Feb.	H	4 n W 1	4 n & nt W 1	6 n W 1
	L	26 n SW 2	26 n SW 2	26 m SW 4
Mar.	H	16 m NNE 1	15 m & n NE 0	15 n SE 0
	L	24 m E 1	10 m W 2	10m&n SW4 W3
Apr.	H	3 m N 2	2 all day NE 1(a)	2 n SE 1(b)
	L	11 m NE 2	30 m S 2	30 m S 1
May	H	13 m NE 2	12 n & nt NE 1(c)	12 all day NE 2(d)
	L	2 m SSW 2	2 m SW 1	2 m NE 0
June	H	21 m&n WSW 1	14 nt NE 0(e)	14nt 15m W0 E0
	L	9 n SSW 2	9 nt SW 0	9 nt calm
July	H	7 n NW 1(f)	7 all day calm	7n&nt S1 NW1
	L	5 n W 1	13 n & nt W 1(g)	20m&n SW3 W2
Aug.	H	18 m WNW 1	11 nt & 12 m W 0	11 alld. NW0(h)
	L	26 m ———	3 n SW 0	3 m & n SW 1
Sep.	H	26 n N 1(i)	26 n N 0	26 n NW 0
	L	3 m W 2	3 m W 0	20 nt SW 3
Oct.	H	16 n E 1	16 n & nt W 0	16n&nt W0 SE0
	L	23 n E 1	12 nt W 2	12 n S 3
Nov.	H	28 m N 1	14 m NE 1(k)	14, 15 all day(l)
	L	21 m SW 2	19 nt N 0	19 all day S 0 E 0
Dec.	H	6 n NW 1	6 n & nt NE 0	6 n calm
	L	18 m W 2	15 m SW 4	15 m WSW 4

(a) And 3 all day E 1. (b) And 3 all day NW 1. (c) And 13 m. NE 1.
(d) And 13 n. NE 2. (e) And 15 m & n. NE 0. (f) And 17 m & n.
NW 1. also 26 m. WSW 1. (g) And 20 m & n. SW 2. (h) And 12 all day
NW 0. also 14 n. & 18 m. SW 0. (i) And 26 m. WNW 1.
(k) And 13 nt & 27 nt. NE 1. (l) E 2. SE 0.

1791.

	KENDAL.			KESWICK.		
	Mean	higheſt	loweſt	Mean	higheſt	loweſt
Jan.	29.33	30.22	28.40	29.23	30.17	28.31
Feb.	29.83	30.47	29.00	29.77	30.42	28.85
March	30.06	30.59	28.88	30.01	30.51	28.82
April	29.72	30.12	28.97	29.66	30.11	28.91
May	29.94	30.37	29.22	29.90	30.37	29.08
June	29.89	30.19	29.50	29.86	30.17	29.40
July	29.76	30.22	29.22	29.71	30.19	29.11
Auguſt	29.96	30.47	29.47	29.91	30.48	29.29
Sept.	30.04	30.34	29.31	30.01	30.31	29.19
Octo.	29.62	30.47	28.56	29.55	30.46	28.45
Nov.	29.58	30.15	28.66	29.51	30.11	28.56
Dec.	29.51	30.28	28.84	29.44	30.19	28.68
Inches	29.77 annual mean			29.71 annual mean		

		KENDAL		KESWICK	
Jan.	H	24 m	W 2	24 m and n	WSW 3 & 4
	L	20 m	E 1	20 m	E 2
Feb.	H	4 nt	calm	4 nt	calm
	L	18 n	NW 1	18 m	W 2
Mar.	H	8 m and n	SW 1, NE 0	8 m and n	E 1
	L	20 nt, 21 m	SW 2	20 nt, 21 m	SW & W 3
Apr.	H	29 nt, 30 m	NE 1	29 nt, 30 m	NW & NE 2
	L	23 m and n	SW 1	23 m and n	NW & N 1
May	H	28 nt	NE 0	28 n	E 1
	L	19 m	W 3	19 m	WSW 4
June	H	7 m	NE 0	7 m	W 1
	L	30 all day	SW 1	30 n and nt	SE 1, SW 0
July	H	15 m and n	NE 0	15 nt	calm
	L	4 n	SW 3	4 m and n	WSW 3
Aug.	H	20 m	NE 0	19 nt, 20 m	calm
	L	28 m	W 1	28 m	W 0
Sep.	H	26 m	NE 0	26 m	NE 0
	L	4 m	SW 0	4 m	SW 1
Oct.	H	29 all day	NE 1	29 m and n	NE 2
	L	21 m	SW 1	21 m	SE 1
Nov.	H	26 nt	S 0	26 nt	SW 2
	L	16 m	SW 1	16 m	SW 1
Dec.	H	17 m and n	NE 0	17 n	NE 0
	L	13 m and n	SW 1	13 m	WSW 3

1792.

	KENDAL.			KESWICK.		
	Mean	higheſt	loweſt	Mean	higheſt	loweſt
Jan.	29.60	30.37	28.87	29.53	30.34	28.65
Feb.	29.87	30.47	29.34	29.82	30.43	29.25
Mar.	29.60	30.41	29.00	29.51	30.40	28.91
April	29.80	30.28	29.16	29.73	30.23	29.02
May	29.88	30.34	29.03	29.82	30.28	28.79
June	29.86	30.37	29.37	29.82	30.39	29.28
July	29.80	30.09	29.47	29.74	30.06	29.37
Auguſt	29.86	30.22	29.12	29.81	30.22	29.02
Sept.	29.65	30.22	29.06	29.59	30.17	28.91
Octo.	29.73	30.47	29.09	29.67	30.45	28.94
Nov.	29.90	30.37	29.09	29.82	30.31	28.96
Dec.	29.71	30.28	28.90	29.62	30.20	28.71
Inches	29.77 annual mean			29.71 annual mean		

		KENDAL		KESWICK	
Jan.	H	5 all day	calm	5 all day	calm
	L	16 m	SW 0	15 nt	S 1
Feb.	H	16 nt, 17 m	NE 1	17 m and n	NE 0
	L	1 m	SW 1	1 m	SW 0
Mar.	H	12 m and n	NE 1	12 m	NE 0
	L	4 n	SW 0	4 n	W 0
Apr.	H	29 m	SW 0	29 m and n	calm
	L	4 nt	SW 1	4 nt	calm
May	H	5 nt, 6m	NE 1	5 nt, 6m	N 2, NE 1
	L	29 m and n	SE 3	29 m	SE 4
June	H	3 all day	N 1	3 n and nt	NE 1, N 0
	L	11 m and n	SW 1	11 m	calm
July	H	15 n, 31 nt	W 0	15 n, 31 nt	SW 1, SE 1
	L	27 m and n	SW 1	27 m	calm
Aug.	H	1m&n 29 all d.	NE0&1	1 n, 29 n	NE 0
	L	23 m	SW 1	23 m	SW 1
Sep.	H	16 m	SW 0	15 nt	SW 0
	L	22 m	SW 0	21 nt	SW 0
Oct.	H	24 nt	NE 0	24 nt	N 1
	L	14 n, 15 m	SW 1	14 n and nt	S 1
Nov.	H	24 nt	NE 0	24 nt	calm
	L	14 n and nt	SE 2	14 n and nt	SE 1
Dec.	H	2 nt	calm	2 nt	calm
	L	6 nt	SW 3	6 nt	SW 4

GENERAL OBSERVATION.

It will be feen from the above accounts, that the barometer is generally higheft and loweft about the fame time
at all the three places; and if the obfervations had been all
taken at the fame hour, it would have been more generally
the cafe.—Whenever the barometer happens to be at the
monthly extreme at one place, and not at another, I find it
is always near it at the other; the greateft differences in this
refpect feem to take place about the lower extreme, and to
be occafioned by rain,—thus, when it happens to be exceffively heavy rain at one place, and not at another, the barometer is relatively loweft where the rain falls.

Mean ftate of the barometer at Kendal *and* Kefwick, *for the
whole 5 years, for each particular month of the year; together
with the means upon the extremes of high and low, and the
mean monthly range.*

	KENDAL.				KESWICK.			
	Mean*	higheft	loweft	range	Mean*	higheft	loweft	range
January	29.68	30.45	28.49	1.96	29.62	30.43	28.41	2.02
February	29.77	30.35	28.99	1.36	29.71	30.33	28.91	1.42
March	29.84	30.36	29.09	1.27	29.77	30.34	29.01	1.33
April	29.79	30.23	29.06	1.17	29.72	30.19	28.96	1.23
May	29.88	30.32	29.27	1.05	29.82	30.27	29.13	1.14
June	29.85	30.26	29.38	.88	29.80	30.24	29.30	.94
July	29.74	30.10	29.40	.70	29.68	30.08	29.31	.77
Auguft	29.86	30.28	29.37	.91	29.80	30.27	29.27	1.00
September	29.80	30.26	29.23	1.03	29.74	30.21	29.12	1.09
October	29.76	30.43	28.99	1.44	29.69	30.40	28.88	1.52
November	29.78	30.31	28.23	1.38	29.70	30.26	28.85	1.41
December	29.72	30.32	28.97	1.35	29.65	30.25	28.84	1.41
Inches	29.79	30.31	29.20	1.21	29.72	30.27	29.00	1.27

The

* The means in this column are corrected, on account of the expanfion
of the mercury, by heat; the correction is made by increafing the height
in the colder months, and leffening it in the warmer months, proportionally
to the defect or excefs of temperature, relative to the mean; it never
exceeds .03 of an inch.

Observations on the Barometer. 17

The mean monthly range at *London*, upon an **average of** the 3 years we have given, is, Jan. 1.73 inches, Feb 1.33, March 0.96, April 0.99, May 0 70, June 0.83, July 0.65, Aug. 0.79, Sept. 1.02, Oct. 0.99, Nov. 1.34, Dec. 1.36. Mean range 1.06 inches.

A Table of the mean spaces described by the mercury each month, determined by summing up the several small spaces ascended and descended; also the mean number of changes from ascent to descent, and the contrary, each month, it being reckoned a change when the space described is upwards of .03 of an inch.—The means are for 5 years, at Kendal *and* Keswick.

	KENDAL.		KESWICK.	
	Mean spaces described by the mercury, in inches.	Mean number of changes, &c.	Mean spaces described by the mercury, inches.	Mean number of changes, &c.
January	9.97	23	10.15	20
February	7.57	21	7.90	20
March	6.64	19	7.30	21
April	6.06	17	6.15	20
May	5.47	19	5.65	19
June	3.89	16	4.25	16
July	4.98	21	5.20	22
August	4.32	18	4.93	19
September	5.87	19	6.59	20
October	6.30	18	6.24	20
November	7.36	18	7.69	20
December	10.08	22	9.95	24
Ann. space	78.51	231	82.00	241

D   SECTION

SECTION SECOND.

Of the Thermometer.

THE next important inſtrument in meteorology is the thermometer : by which the
temperature, or degree of heat, of the air and
other bodies, is determined. An inſtrument
under this character was invented prior to the
barometer, but never brought to a tolerable degree of perfection till the preſent century.

Philoſophers are generally perſuaded, that the
ſenſations of *heat* and *cold* are occaſioned by the
preſence or abſence, in degree, of a certain principle or quality denominated *fire*, or *heat ;*—thus,
when any ſubſtance feels cold, it is concluded
the principle of heat is not ſo abundant in that
ſubſtance as in the hand . and if it feel hot, then
more abundant. It is moſt probable, that all
ſubſtances whatever contain more or leſs of this
principle. Reſpecting the nature of the principle, however, there is a diverſity of ſentiment :
ſome ſuppoſing it a *ſubſtance*, others a *quality*,
or property of ſubſtance. *Boerhaave*, followed
by moſt of the moderns, is of the former opinion ; *Newton*, with ſome others, are of the
latter ; theſe conceive heat to conſiſt in an internal vibratory motion of the particles of bodies.
Whatever

Whatever doubts may be entertained refpecting the *caufe* of heat, many of its effects are
clearly afcertained : in treating of thofe effects
it is expedient to adapt our language to one or
other of the fuppofitions refpecting the nature of
their caufe ; and as nothing has yet appeared to
render the common mode of expreffion unphi-
lofophical, we fhall therefore fpeak of fire as
a *fubftance*, under the denomination of fire,
or heat.

One univerfal effect of fire is its expanding or
enlarging thofe bodies into which it enters ;
which bodies fubfide again when the fire is
withdrawn. *Solids* are leaft expanded by it ;
inelaftic fluids, as water, fpirits, &c. are more
expanded ; and *elaftic fluids*, as air, moft of all.
Hence, if a glafs tube of very fmall bore, and a
large bulb at the end, be filled with any liquid
fo as it may rife into the ftem, and heat be applied to the bulb, the liquor will rife in the tube,
and it is obvious to infer, that the larger the
bulb and the fmaller the bore of the tube, all
other circumftances being the fame, the greater
will be the afcent for a given variation of heat :
fuch an inftrument, when applied to a frame
properly graduated, is called a *thermometer.*
Different fluids have been occafionally ufed for
thermometers, but none is found to anfwer fo
well, in all refpects, as quickfilver.

Boiling

Boiling water is of a conftant and uniform temperature at all times and places, provided the barometer be at a certain height; and a mixture of pounded ice, or fnow, and water is likewife of an uniform temperature. Hence, we are favoured with the means of finding two fufficiently diftant points upon the thermometric fcale, without the neceffity of another thermometer; thefe are called the *boiling* and *freezing points*, and are marked with 212° and 32° refpectively, upon the common fcale, or that of *Fahrenheit*, the boiling point being found when the barometer ftands at 30 inches. The fcale is divided into equal parts, and extended above and below thefe points, *ad libitum;* when the degrees go below o, they are counted from it, and termed *negative*, merely for diftinction*. At 55° the word *temperate* is ufually placed upon the fcale, and *fummer heat* at 75°; 98° denotes the ufual heat of the human blood; 112° the heat of the blood fometimes in an inflammatory fever; and at 175° fpirits boil: quickfilver itfelf boils at about 600°.

Reaumur's fcale, ufed by fome philofophers on the continent, marks the freezing point with o°, and the boiling point with 80°.

THE

---

* The Tyro will pleafe to obferve, that the term o°, does not imply a total deprivation of fire; it is a mere arbitrary term, and there would have been no lefs propriety in calling it 100°, or 1000°, than o°.

THE following is the refult of obfervations on the thermometer, taken three times a day, at *Kendal* and *Kefwick*,
from 1788 to 1792, inclufive. The morning obfervations
were taken between 6 and 8 o'clock ; the mid-day obfervations about 12 or 1 ; the night obfervations at *Kendal* about
9 or 10, but at *Kefwick* at 6 in fummer, and 4 in winter ;
this circumftance makes the mean temperature of *Kefwick*
too high, when compared with that of *Kendal,* which ought
to be noticed in the comparifon.

The fituation of the thermometers too, is another particular that fhould be adverted to ;—that at *Kendal* was
without, in a garden, under the fhade of a pretty large
goofeberry tree, facing the north: the garden is open to
the country in the north, and has houfes at the diftance of
8 or 10 yards to the fouth. The thermometer at *Kefwick*
is fituate near, but not in contact with, the wall and window
of a houfe facing the north, which is open to the conntry :
it is about 6 yards above the ground ; the fun never fhines
on it in winter, and only a few weeks in fummer, and that
early in the morning, long before the obfervation is taken.

From thefe accounts it is obvious to infer, that the thermometer at *Kefwick* will not be liable to the *extremes* of heat
and cold, owing to the influence of the adjoining wall ;
whereas that at *Kendal* is perhaps liable to too great an extreme of heat, occafionally, owing to the reflection from the
ground, though the fun never fhines upon the frame for an
hour at leaft before any obfervation is taken.

The following tables, it is prefumed, will be fufficiently
explicit ; we have given a table each year, containing the
days on which the extremes of heat and cold happened at
each place, as with the barometer.

AT

AT KENDAL, 1788.

	Morning.			Noon.			Night.			month.
	Mean	high.	low.	Mean	high.	low.	Mean	high.	low.	means.
	°	°	°	°	°	°	°	°	°	°
Jan.	37.7	46	20	41	47	31	38.3	46	24	39
Feb.	36.3	44	28	41	47	31	37.7	46	28	38.3
March	33.9	46	18	40.3	50	31	36.3	50	21	36.8
April	43.8	49	32	49.5	69	39	45 8	55	34	46.3
May	48.7	61	39	61.8	80	43	48 6	61	38	53.0
June	55.0	60	47	66.4	80	57	52 5	60	45	57.3
July	55.0	62	49	61.0	68	53	54.4	62	47	56.8
Aug.	53.5	58	47	63.7	74	57	54.2	60	49	57.1
Sep.	49.5	60	35	59.4	70	50	51.8	62	43	53.6
Oct.	41.5	55	28	52.6	58	47	43.1	57	30	45.7
Nov.	38.3	50	27	44.5	52	33	39.3	52	28	40.7
Dec.	26	40	10	33.5	46	23	27.6	40	18	29
an. m.	43.1			51.2			44.1			46.1

AT KESWICK, 1788.

(The obſervations on the thermometer at Keſwick this year were not complete till May)

	Morning.			Noon.			Night.			month.
	Mean	high.	low.	Mean	high.	low	Mean	high.	low.	means.
	°	°	°	°	°	°	°	°	°	°
May	54.8	71	41	61	77	41	58.9	72	39	58.2
June	57.6	67	52	62.3	77	56	60.5	75	51	60.5
July	58.6	68	52	61.7	70	58	60.3	64	56	60.2
Aug.	58.5	62	52	63.2	75	56	61.6	72	53	61.1
Sep.	54.6	64	46	58.3	68	48	55.9	66	48	56.3
Oct.	44.4	57	28.	50.5	59	41	46.8	57	34	47.2
Nov.	42.1	52	28	44.2	52	32	42.5	54	28	42.9
Dec.	26.2	43	8	30.2	44	18	28.5	41	17	28.3

1788.

1788.

		Morning		Noon		Night	
		Kendal.	*Keswick.*	*Kendal.*	*Keswick.*	*Kendal.*	*Keswick.*
Jan.	H	24th day		24th & 27th days		26th day	
	L	15		15		24	
Feb.	H	15		15		14	
	L	2		2		10	
Mar.	H	30		22, 30		30	
	L	7, 11		13		7, 8	
Apr.	H	11, 12		30		30	
	L	4. 5		4		4	
May	H	28	26	26	26, 27	27	25, 27
	L	6	29	29	29	10	29
June	H	18	16, 17	21		17	17, 18, 20 · 17
	L	1, 8	12, 20, 28	19	27, 28, 30	9	19
July	H	12		13 · 12		13	11 · 12, 13, 30
	L	28		5 · 10	7, 8, 22	8	25
Aug.	H	1, 13 · 3, 10, 13	14 · 4		4	12 · 3, 4	
	L	18	27 · 23		26	19 · 24	
Sep.	H	5	8 · 4	4, 8 · 4		4	
	L	15	21 · 29	21 · 14		21	
Oct.	H	2	2 · 2. 5, 22	2, 22 · 2		22	
	L	19	18 · 18, 19, 20	18 · 18		18	
Nov.	H	11, 12	11 · 3, 11	2, 3, 11, 12 · 12		3, 11	
	L	16	15 · 16	15 · 15		26	
Dec.	H	24	25 · 24	24 · 24		24	
	L	16	8, 16	15, 16 · 15, 17, 28		17, 28	

AT KENDAL, 1789.

	Morning			Noon			Night			month means.
	Mean	high.	low.	Mean	high.	low.	Mean	high.	low.	
	°	°	°	°	°	°	°	°	°	°
Jan.	30.8	47	4	36	50	19	32	47	5	33
Feb.	37.7	46	28	42.2	47	32	37.7	46	30	39.2
March	30.5	37	22	41.4	48	31	32	37	25	34 6
April	39 4	47	25	49.8	60	36	40.5	49	31	43.2
May	48.8	55	41	60	70	51	49.4	57	38	52.7
June	51 5	60	38	63.6	79	53	51.6	60	45	55 6
July	54	62	41	64.8	74	54	54.4	62	49	57.7
August	53.7	62	42	69.4	79	60	56	62	50	59.7
Sept.	49.8	57	32	58.4	67	50	49.2	58	37	52.5
Octo.	42.4	55	27	51	57	39	43.4	55	29	45.6
Nov.	35	44	20	42.5	51	31	35.7	45	23	37.7
Dec.	40.6	48	25	43.4	49	35	40.9	49	29	41.6
ann. ms.	42.8			51.9			43.6			46.1

AT KESWICK, 1789.

	Morning.			Noon.			Night.			month.
	Mean	high.	low.	Mean	high.	low.	Mean	high.	low.	means.
	°	°	°	°	°	°	°	°	°	°
Jan.	33	49	7	34.9	50	20	33.8	50	16	33.9
Feb.	37.7	46	29	40.4	48	34	38.7	47	30	38.9
Mar.	29.7	39	23	36.4	46	29	33.2	40	28	33.1
April	41.6	49	28	45.9	56	33	44.1	51	29	43 9
May	50.9	60	41	56 7	68	45	54	68	41	53.9
June	55.3	69	44	59	73	49	55.8	69	47	56.7
July	58.2	66	49	62.4	71	51	60.2	70	50	60.3
Auguſt	59.6	67	51	64.9	74	67	62 9	71	51	62.5
Sept.	52.7	61	41	56.4	65	46	54.8	64	46	54.6
Octo.	44.2	52	26	48.9	56	36	46.9	57	32	46.7
Nov.	37	45	23	41.2	48	31	39.1	48	28	39.1
Dec.	42.5	49	31	43.4	49	34	42.2	52	34	42.7
ann. ms.	45 2			49 2			47.1			47.2

1789.

		Morning.		Noon.		Night.	
		Kendal.	Keſwick.	Kendal.	Keſwick.	Kendal.	Keſwick.
Jan.	H	30	30	31	30	26	30
	L	12	12	5, 12	7	5	11, 12
Feb.	H	15	1, 15	1, 15, 18, 24	15	14	1
	L	12	27	4	11, 12, 16	7	11
Mar.	H	20, 21	13	21	21	2, 28	2
	L	24	8, 17	13	7	11	7, 10, 13
Apr.	H	21	20, 21	30	30	16	16, 30
	L	12	4	7	3	4	3
May	H	24, 25, 28	26, 28	13	13	13	13
	L	1, 11	2, 4	5	18	3	3
June	H	18	18	17	17	17	13, 16
	L	2	26	6, 26	6, 28	7, 25	27
July	H	4	3	4	4	4	3
	L	24	23	22	22	23	22
Aug.	H	5	4	13	18	4	18
	L	23	1	7, 22	6	22	30
Sep.	H	1, 4, 9	1	5	6	3, 10	1
	L	17	17	15, 19	19	18	16
Oct.	H	21	21	23	23	20	20
	L	26	31	31	31	31	31
Nov.	H	14	2, 14	3	2	1	10
	L	27	26	27	26	26, 27, 28	26
Dec.	H	6	22	6, 27	6, 7	23, 27	7
	L	1	16	16	16	25	17

AT KENDAL, 1790.

	Morning.			Noon.			Night.			month.
	Mean	high.	low.	Mean	high.	low.	Mean	high.	low.	means.
	°	°	°	°	°	°	°	°	°	°
Jan.	36.3	47	22	40	49	31	37.2	47	24	37.8
Feb.	39.8	47	25	46.2	54	39	41.2	47	28	42.4
March	35.5	45	24	50.7	58	39	38.1	46	29	41.4
April	37.5	46	23	50.1	58	40	37.5	47	28	41.7
May	48.5	55	38	59.8	71	51	48	58	43	52.1
June	52.4	62	43	61.7	76	53	53.4	62	46	55.8
July	51.6	58	41	60.9	67	54	53.2	59	47	55.2
Aug.	52.5	60	38	61.4	71	56	55	63	48	56.3
Sep.	47	56	33	56.7	68	51	49.1	52	42	50.9
Oct.	43.3	55	28	54 3	65	47	45.4	58	33	47.6
Nov.	37.8	50	24	44.1	53	34	37.6	48	21	39.8
Dec.	34.3	46	6	39	49	22	34.9	47	22	36.1
an. m.	43.0			52.1			44.2			46.4

AT KESWICK, 1790.

	Morning.			Noon.			Night.			month.
	Mean	high.	low.	Mean	high.	low	Mean	high.	low.	means.
	°	°	°	°	°	°	°	°	°	°
Jan.	36.4	48	29	39.7	50	30	38.7	50	32	38.6
Feb.	43.3	49	32	45.4	51	39	43.8	49	35	44.2
March	39.3	49	31	47.7	52	38	44.3	50	38	43 8
April	39	49	31	44.8	54	38	41.2	49	32	41.7
May	50.8	58	45	55.5	64	46	52.9	63	45	53.1
June	55	68	47	58.7	71	49	57.6	70	49	57.1
July	53.9	59	48	59.5	65	52	56.5	63	51	56.6
Aug.	54.8	61	51	60.5	69	55	57.5	64	51	57.6
Sep.	49.2	56	41	53.7	62	43	50.8	62	43	51.2
Oct.	46.5	59	35	52.1	62	44	48.8	59	37	49.1
Nov.	37.6	49	25	41.6	50	30	39.9	50	26	39.7
Dec.	35.7	47	18	37.9	49	27	36.8	47	28	36.8
an. m.	45.1			49.8			47.4			47.4

1790.

		Morning		Noon		Night	
		Kendal.	Keswick.	Kendal.	Keswick.	Kendal.	Keswick
Jan.	H	3, 12	14	3	12	12	12
	L	21	9, 15, 21, 30	15, 17	15	20	20, 25
Feb.	H	25	26	28	22	24	22, 25
	L	21	1	1	10	20	10
Mar.	H	2, 12	2	28, 30	20, 22, 30	11	18
	L	17	4, 15, 17, 18	5, 10	10	15, 16	5, 15
Apr.	H	23, 27	29	19	23	22	28
	L	17	11, 13	16	13	14	12, 13, 14
May	H	30, 31	30	29, 31	31	29	16
	L	14	19, 21	4, 27	2	3	5
June	H	22	16	15	22	15, 16	22
	L	13, 28	5, 10	11	11	11	6, 7
July	H	26	4, 26	27	17, 25	17	24
	L	30	30	29	31	21	14
Aug.	H	16	21	12	15	7	16
	L	27	1. 3, 4, 26, 27	4. 9	3, 23	26	29
Sep.	H	12	19	19	19	12	19
	L	8	15	21	14	7	14
Oct.	H	4, 5, 21	22	4	6	21	22
	L	10	25	27	30	9, 30	30
Nov.	H	6	6	6	6	25	6
	L	29	28	18, 29	27	30	30
Dec.	H	10, 13	13	13	13	9	7
	L	20	20	20	1, 20, 28	28	28

AT KENDAL, 1791.

	Morning			Noon			Night			month, means.
	Mean	high.	low.	Mean	high.	low.	Mean	high.	low.	
	°	°	°	°	°	°	°	°	°	°
Jan.	38.3	48	23	40.4	48	32	38.8	48	30	39.2
Feb.	36.5	46	26	41.9	50	36	36.5	46	28	38.3
March	38.3	47	23	47.2	55	39	40.5	48	25	42
April	43.7	53	36	52.8	67	42	44.1	55	37	46.9
May	45	55	34	56.5	73	44	45.1	61	33	48.9
June	51.1	59	38	63.8	81	48	52.2	62	40	55.7
July	54.3	67	48	63.5	78	51	54	66	48	57.3
August	54.8	66	45	64.3	74	48	53.8	62	46	57.6
Sept.	50.3	60	38	63.4	79	52	51.5	60	42	55.4
Octo.	43	57	24	51.8	60	42	43.9	57	24	46.2
Nov.	39.4	49	22	45.2	53	39	39.3	50	28	41.3
Dec.	29	40	10	33	42	20	30.2	46	-10*	30.7
ann. ms.	43.6			52			44.2			46.6

* Thermometer at 8½ P. M. —6°; at 9¼, —10°, and till 10 P. M.—10°.

AT KESWICK, 1791.

	Morning.			Noon.			Night.			month. means.
	Mean	high.	low.	Mean	high.	low.	Mean	high.	low.	
Jan.	37 9	45	24	39 3	49	31	39.2	49	31	38.8
Feb.	35 8	47	25	38.4	47	30	36.7	47	30	37
Mar.	40.4	47	22	43 6	54	34	41.8	48	29	41.9
April	44.4	54	36	49.6	65	41	46 1	59	37	46.7
May	47	64	37	52 7	70	40	48.9	66	40	49.5
June	54 2	70	41	59 3	76	45	56 3	73	41	56.6
July	56 4	70	50	59.8	73	51	57 6	71	50	57.9
Auguſt	55.6	64	48	60 9	68	47	58 2	67	45	58.2
Sept.	53.2	66	41	59.1	73	49	56.6	69	46	56 3
Octo.	43 5	55	27	47.7	61	36	46.4	58	32	45.9
Nov.	39 1	47	22	42.3	51	33	40.6	51	32	40.7
Dec.	29.9	39	13	33.2	40	22	32.1	41	15*	31.7
ann. ms.	44 8			48 8			46.7			46.8

1791.

		Morning.		Noon.		Night.	
		Kendal.	Keſwick.	Kendal.	Keſwick.	Kendal.	Keſwick
Jan.	H	31	25	16, 31	16	16, 30	10
	L	28	28	5	28	3	28
Feb.	H	14	14	7	14	10	14
	L	4	4	4, 23	4	16, 28	2. 3
Mar.	H	15	29, 30	28, 30	30	22	29
	L	2	1	3	21	1	1
Apr.	H	17	15	16	16	16	16
	L	1	11	6	5	25	6, 10
May	H	28	31	30	30	30	30
	L	4, 8	6	1	1	6	3, 6, 23
June	H	4	4, 5	6	4	3, 5	3
	L	14	11, 12	21	12	13, 21	12
July	H	18	17	17	16, 17	17	17
	L	5, 14	5, 6	4	4	10	4
Aug.	H	15	23, 24	15	12, 20, 23	14	23
	L	19	30, 31	31	31	31	31
Sep.	H	10	11	11	11	10	11
	L	30	19	22	19	1	18
Oct.	H	4	4	3, 4, 5	3	3	3, 4
	L	24	23	26	22	23	23
Nov.	H	11	11	13	11	11	11
	L	6	6	6, 18, 30	6	6	18
Dec.	H	2, 31	23, 31	2, 31	1, 27, 31	1	1
	L	15	11	11	11	11	11

* Thermometer at 4 P. M. 15°; at 10 P. M. 8°; at 1 A. M. 6°.

AT KENDAL, 1792.

	Morning.			Noon.			Night.			month. means.
	Mean	high.	low.	Mean	high.	low.	Mean	high.	low.	
	°	°	°	°	°	°	°	°	°	°
Jan.	32.3	46	11	37.5	48	26	32.5	49	12	34.1
Feb.	37	49	18	44	55	35	37.6	46	27	39.5
March	38.2	48	20	46.3	54	32	39.3	48	22	41.2
April	45	52	36	54.4	72	43	43.9	50	29	47.8
May	45.3	52	35	54.8	62	47	45.4	52	36	48.5
June	52.1	61	48	61.4	75	52	50.2	58	44	54.5
July	56.1	62	51	64.3	72	56	54.9	62	5c	58.4
Aug.	55.6	66	48	69	83	58	56.1	66	48	60.2
Sept.	47.6	59	34	56.8	69	46	48.5	59	36	51
Oct.	43.4	58	28	51.4	63	46	44	56	35	46.3
Nov.	41.6	50	24	47.1	58	34	42	51	30	43.6
Dec.	36.9	50	21	40.3	51	29	37.7	52	24	38.3
an. ms.	44.3			52.3			44.3			47

AT KESWICK, 1792.

	Morning.			Noon.			Night.			month. means.
	Mean	high.	low.	Mean	high.	low.	Mean	high.	low.	
	°	°	°	°	°	°	°	°	°	°
Jan.	32.1	45	12	35	47	20	33.4	47	17	33.5
Feb.	36.5	47	19	41.2	53	30	39.7	50	30	39.1
March	37.3	47	17	42.3	51	25	39.6	47	21	39.7
April	44.8	53	35	51.2	67	41	46.9	65	35	47.6
May	46.5	53	36	51.8	62	40	48.2	56	39	48.8
June	53.4	63	46	57.1	66	48	54	63	46	54.8
July	57.3	63	49	61.4	69	50	57.9	65	48	58.9
Aug.	60	71	47	65.2	75	54	59.9	70	48	61.7
Sept.	48.9	59	38	53.9	64	43	51.2	60	42	51.3
Oct.	44.9	57	32	48.8	64	41	46.6	57	37	46.7
Nov.	42.2	55	23	45.7	60	32	44.2	57	33	44
Dec.	36.4	51	20	38.7	51	23	38	49	24	37.7
an. ms.	45.0			49.4			46.6			47

1792.

1792.

		Morning.		Noon.		Night.	
		Kendal.	Kefwick.	Kendal.	Kefwick.	Kendal.	Kefwick.
Jan.	H	31	31	30, 31	31	30	30
	L	13	11	11	11, 12	11	11
Feb.	H	2	2	26	12	6, 27, 29	26
	L	21	21	18, 19, 24	18	20, 21	18, 20, 21
Mar.	H	29	1	17	1, 2	17, 24, 28	2, 17, 30
	L	11	9	13	8	9	8
Apr.	H	13, 14, 29	12, 13	10	11	11, 13	11
	L	7	5, 20	5	5	20	19
May	H	19	15	27	12	24, 27	24
	L	11	1	1, 3	1, 2	2	1, 10
June	H	5	16	29	16	16	4
	L	2, 10, 20	8, 10	12	19	2	19
July	H	16	9, 10, 15, 24	29	29	15	31
	L	30	1	5, 11	11	28	11
Aug.	H	1	3	1	3	2	3
	L	28, 29	28	20	28	19, 28	28
Sep.	H	13	2, 7	2	2	7	4
	L	16, 22	22	21	22	15	27
Oct.	H	1	1	1	1	1	1
	L	12	24	12	3, 4	11	24
Nov.	H	3, 5, 8, 11	5	5	5	3	5
	L	17	17	20	17	19	16, 17, 19
Dec.	H	10, 18, 20	18	18	18	18	18
	L	25, 31	24	23	24	7	23

The monthly and annual means of the thermometer, upon 5 years, are as under.

	Jan.	Feb.	Mar.	April	May	June	July	Aug.
At Kendal	36.6	39.5	39.2	45.2	51	55.8	57.1	58.2
At Kefwick	36.8	39.5	39	45.3	52.7	57.1	58.8	60.2

	Sept.	Oct.	Nov.	Dec.	annual mean.
At Kendal	52.7	46.3	40.6	35.1	46.4
At Kefwick	53.9	47.1	41.3	35.4	47.3

The

The annual mean at *Kefwick* may perhaps be ftated more accurately at 46°, as the evening obfervations were taken too foon to give the true mean temperature. It may however be proper to obferve here, that the time or times of the day at which the obfervations ought to be made, in order to determine the true mean, has not, that I know of, been afcertained.

I made the following obfervations on the temperature of a pump well, the furface of which is ufually from 3 to 6 feet below that of the ground; at the end of January its heat was 45°; February, 45°; March, 46°; April, 46°.5; May, 48°; June, 50°; July, 51°; Auguft, 52°; September, 50°; October, 48°.5; November, 47°.5; December, 45°.————Thefe obfervations give an annual mean of 47°.8.

About the middle of June, 1793, I found the temperature of feveral wells in *Kendal*, after having pumped a few gallons of water from each; fix of the deepeft, being from 5 to 10 yards below the furface of the ground, were juft 48° each; three other deep ones were 47°.5; one not quite fo deep was 46°; and three that were only 2 or 3 yards below the furface of the ground were 49° each. The deep ones, I believe, in general are fubject to very little variation in temperature all the year round.

From thefe obfervations on the temperature of wells, I am inclined to think, the heat of the earth at 10 or more yards depth is not the fame, at *Kendal*, as the mean heat of the air, but fomething greater. Perhaps this is a general fact; the temperature of the cave of the obfervatory at *Paris*, which is 30 yards below the pavement, is 53°.5; whereas the mean heat of the air there, is only 52°.——However this may be, I cannot believe the mean heat of the air at this place is fo great as that of the pump water.

SECTION

SECTION THIRD.

Of the Hygrometer.

THE hygrometer is an inftrument meant to fhew the difpofition of the air for attracting water, or for depofiting the water it has in folution with it.

Some of the greateft philofophers of the prefent age have been endeavouring to improve thofe inftruments of this defcription we have already, and to invent others lefs objectionable; but I prefume the object is not yet fully attained. —To afcertain the exact quantity of water in a given quantity of air, and alfo the difpofition of the air for imbibing or depofiting it, is an object indeed, highly important to the fcience of meteorology, and to philofophy in general.

It does not fuit our intended brevity to enter into a detail of the different inftruments lately propofed, with their refpective merits and demerits; we fhall only obferve, that moft fubftances are affected more or lefs with the drynefs and moifture of the air, particularly animal and vegetable fibres, which become turgid, and contract by being expofed to moift air. Spunge, paper, &c. imbibe moifture, and become alternately

nately lighter and heavier by being expofed to the air. Strings, whether made of animal or vegetable fibres, twift and untwift by the moifture and drynefs of the air, and confequently are fhortened and lengthened alternately.——— The force with which a cord contracts is amazingly great. Mr. *Boyle,* who feems to have been the firft that made a feries of experiments of this fort, ufed to fufpend a weight of 50 or 100lbs. to the end of a rope, which was alternately raifed and lowered by the moifture and drynefs of the air, as a fmall weight would have been.

Obfervations on the Hygrometer.

THE only hygrometrical inftrument I have ufed, is a piece of whip-cord, about 6 yards long, faftened to a nail at one end, and thrown over a fmall pulley ; in this manner it has been kept ftretched, by a weight of 2 or 3 ounces, fince September, 1787. It is in a room without a fire, and where the air has a moderate circulation; the fcale is divided into tenths of an inch, and begins at no determined point; the greater the number of the fcale, the longer is the ftring, and the drier the air. This ftring has varied in length above 13 inches, or $\frac{1}{16}$ of its whole length. The obfervations were taken three times a day the two firft years, and once a day after, namely, at noon.———The refult follows.

Mean

Mean ftate of the Hygrometer, at Kendal.

	1788.	1789.	1790.	1791.	1792.	Mean of the whole
January	40.3	83.4	85	85	102.6	79.3
February	54.7	81	92.8	97	100.5	85.2
March	81	106	109	105.7	112	102.7
April	85	112	131.7	116	125	113.9
May	116	123	129	123	128	123.8
June	127	127	129	135	137	131
July	104	126	126	131	134	124.2
Auguft	113	132	121	129.6	138.5	126.8
September	108.5	114	117	129	120.7	117.2
October	102	104	109	119	123.3	111.5
November	87.6	99	104	113	106.4	102
December	100	85	92	107.7	102.6	97.5
An. means	93.3	107.7	112.1	115.9	119.2	
Drieft	138	140	141	144	150	
Moifteft	15	63	71	65	83	

It is obvious, from the means of the feveral years, and likewife from the extremes, that the cord has been increafing in length each year, fo that, in fimilar ftates of the air, the index pointed at greater numbers each year fucceffively; this increafe too appears to have been nearly in arithmetical pro-greffion after the firft year.—In confequence of this increafe in the length of the cord, fome allowance ought to be made in comparing the mean ftate of the hygrometer in the different months of the year; thus, if the months of June or July be taken for a ftandard of comparifon, then the means of the preceding months muft be increafed, and thofe of the following diminifhed, in fuch proportion as the annual in-creafe fhall require.

The above mentioned inftrument ferves to fhew a varia-tion in the drynefs or moifture of the air; but it is very inadequate to the purpofe for which a hygrometer is defired.

F   SECTION

SECTION FOURTH.

Of Rain-gauges, and an account of the quantity of rain that fell at Kendal *and* Kefwick, *in the years* 1788, 1789, 1790, 1791, *and* 1792, *together with the quantity at* London *in the three firſt of theſe years.*

THE *rain-gauge* is a veſſel placed to receive the falling rain, with a view to afcertain the exaɛt quantity that falls upon a given horizontal furface at the place. A ſtrong funnel, made of ſheet iron, tinned and painted, with a perpendicular rim two or three inches high, fixed horizontally in a convenient frame with a bottle under it to receive the rain, is all the inſtrument required.

In order to determine the depth of water that falls in the open field, with this apparatus, we muſt have given, 1ſt. the weight of the water caught in the bottle; 2d, the area of the aperture of the funnel; and, 3d, the weight of a cubic foot of water, which has been found equal to $62\frac{1}{2}$lbs. avoirdupoiſe. Then, if $a =$ the area of the aperture. in inches, $W = 62\frac{1}{2}$lbs. and $w =$ the weight of the water caught, in pounds, we ſhall have this theorem, per menſuration,

20736

$\dfrac{20736w}{aW}$ = the depth of water, in inches, that falls upon any horizontal furface at the time and place, as required.

By inverting this theorem, one may eafily find the weight of water correfponding to any given depth ; which being once found, it is moft expeditious, and fufficiently accurate, when the funnel has 8 or 10 inches diameter, not to *weigh* the water each day, but to *meafure* it, by means of phials, &c. fuitable for the purpofe.

IN the following account, we have given the amount of the rain each month, at *Kendal* and *Kefwick*, for 5 years, except for 3 months at the laft place ; and alfo at *London,* for 3 years : the laft is taken from the Philofophical Tranfactions. The rain at the two before mentioned places was taken each evening at 8 or 10 o'clock.——To the account we have added, the number of *wet days* each month, or thofe on which the rain amounted at leaft to 001 of an inch.

N. B. My rain-gauge at *Kendal* is 10 inches diameter ; and Mr. *Crofthwaite*'s at *Kefwick* about 8 : they were both fufficiently diftant from trees, houfes, &c.

1788.

	At Kendal.		At Kefwick.		At London.
	Inches of rain.	wet days	Inches of rain	wet days	Inches of rain.
Jan.	5.6160	20			0.439
Feb.	3.3064	23			1.461
March	2.8183	16			0.336
April	2.9047	16	3.9204	22	0.607
May	1.1872	10	2.0840	9	0.497
June	2.3137	7	3.6876	9	3.275
July	7.0323	28	6.3757	28	1.620
Aug.	3.0883	18	5.0771	19	2.699
Sept.	4.6756	19	7.1382	23	3.345
Oct.	2.1220	11	1.7537	13	0.103
Nov.	3.0460	18	3.2841	17	0.510
Dec.	1.1470	7	0.9849	12	———
Total	39.2575	193	34.3057	152	14.892
from Mar.	27.5168	134			

1789.

	At Kendal.		At Kefwick.		At London.
	Inches of rain.	wet days	Inches of rain.	wet days	Inches of rain.
Jan.	7.343	22	8.5435	26	1.345
Feb.	8.924	24	9.0442	27	1.605
March	1.347	15	1.3245	21	1.549
April	4.778	19	4.2383	21	0.957
May	5.388	20	3.6611	25	1.103
June	4.311	18	7.0637	19	3.244
July	6.389	25	5.2770	26	2.467
Aug.	1.556	12	3.4569	14	1.864
Sept.	5.436	24	7.2709	24	2.155
Oct.	6.864	21	8.0907	25	3.253
Nov.	5.451	16	6.0965	21	1.244
Dec.	12.048	28	8.1776	27	1.190
Total	69.835	244	72.2449	276	21.976

1790.

1790.

	At Kendal.		At Keſwick.		At London.
	Inches of rain.	wet days	Inches of rain.	wet days	Inches of rain.
Jan.	6 567	18	5.9377	19	0.967
Feb.	3 662	15	4.0124	17	0.115
March	1.606	10	1.3228	10	0.122
April	1.960	11	2.3198	17	1.470
May	2.645	14	3.4588	18	2.898
June	4.114	17	5.1077	21	0.708
July	7.894	25	6.2509	24	1.700
Aug.	6.200	26	5.8524	26	1.991
Sep.	6.682	16	8.3950	20	0.368
Oct.	5.382	15	6.1304	16	1.108
Nov.	5.345	12	5.0550	13	2.512
Dec.	10.306	24	10.9010	24	2.093
Total	66.263	203	64.7439	225	16.052

1791. | 1792.

	At Kendal.		At Keſwick.		At Kendal.		At Keſwick.	
	Inches of rain.	wet days.	Inches of rain.	wet days.	Inches of rain.	wet days.	Inches of rain.	wet days
Jan.	8.369	28	11.3574	28	4.120	13	4.5041	15
Feb	6.641	16	9.2244	21	5.820	14	4.9375	20
March	3.641	17	3.1231	17	6.684	23	9.6261	26
April	4.810	17	3.3190	21	10.091	16	11.6460	17
May	3.983	18	3.9963	18	5.922	19	6.5167	21
June	3.493	13	2.0133	20	3.514	16	2.7110	20
July	6.344	18	8.2060	20	5.926	21	3.8643	20
Aug.	5.165	17	5.8852	16	7.398	18	5.9704	16
Sep.	3.409	10	2.7715	11	11.229	28	10.6179	25
Oct.	5.505	22	7 1272	23	6.028	20	6.7357	21
Nov.	6.465	21	8.7238	23	6.030	18	5.8350	14
Dec.	8.375	22	7.8050	23	12.122	27	11.6404	23
Total	62.200	219	73.5522	241	84.884	233	84.6051	238

Mean

Mean monthly rain and number of wet days, at Kendal *and* Kefwick, *for all the* 5 *years.*

	At Kendal.		At Kefwick.	
	Inches of rain.	wet days.	Inches of rain.	wet days.
Jan.	6.403	20	7.3558	22
Feb.	5.671	18	6.1624	22
March	3.219	16	3.7324	18
April	4.909	16	5.0887	20
May	3.825	16	3.9434	18
June	3.549	14	4.1167	18
July	6.717	23	5.9948	24
Auguſt	4.681	18	5.2484	18
Sept.	6.286	19	7.2387	21
Octo.	5.179	18	5.9675	20
Nov.	5.267	17	5.7989	18
Dec.	8.800	22	7.9018	22
Total	64.506	217	68.5495	241

The greateſt quantity of rain in 24 hours, for theſe 5 years, was on the 22d of April, 1792, namely, at *Kendal*, 4.592 inches. The rain at *Kcfwick* on that day, was fomething lefs; but taking both the 22d and 23d, the rain was nearly equal at both places.

Befides theſe 2, there were other 2 days, at *Kendal*, when the rain was betwixt 2 and 3 inches, and 56 days betwixt 1 and 2 inches.

At *Kefwick*, for 4 years and 9 months, there were 3 days, befides the 2 above mentioned, when the rain was between 2 and 3 inches, and 52 days between 1 and 2 inches.

SECTION

SECTION FIFTH.

Obfervations on the Height of the Clouds.

MR. *Crofthwaite*, of *Kefwick*, has availed himfelf of his fituation in the vicinity of high mountains, to make obfervations on the height of the clouds; for which purpofe he has chofen *Skiddaw*, the higheft mountain in the neighbourhood, a very fine view of which his *mufeum* commands. By means of marks on the fide of the mountain, and with the affiftance of a telefcope, he can difcern, to a few yards, the height of the clouds, when they are below the fummit, which is very often the cafe.—Perhaps the following feries of obfervations is the only one of the kind extant, as the labour and difficulty attending fuch obfervations in a champaign, or flat country, are fufficient to deter any one from making two or three daily obfervations for a feries of years; and when the whole fky is clouded, they are quite impracticable.

He has determined, by trigonometry, the perpendicular height of *Skiddaw*, above the level of *Derwent lake*, to be 1050 yards, which agrees very nearly with Mr. *Donald*'s obfervations; and he has noted, in a column of his meteorological journal,

journal, every morning, noon, and evening, the height of the clouds, in yards, above the level of the ſaid lake, when their height did not exceed that of *Skiddaw;* and when it did, he has marked it as ſuch.

The reſult of 5 years obſervations is contained in the following table. All the obſervations when the clouds were between o and 100 yards high are placed in one column, and thoſe when they were between 100 and 200 yards high in the next column, &c.—In order to determine what effect the ſeaſons of the year have upon the clouds, in this reſpect, we have kept the obſervations in the ſeveral months diſtinct.—It is to be noted, that the column containing the number of obſervations when the clouds were above *Skiddaw,* includes thoſe obſervations when there were no clouds viſible; but Mr. *Croſthwaite* has noted this laſt circumſtance alſo, in the journal, and it appears, that about 1 obſervation in 30, of thoſe in that column, ſhould be deducted on that account.

Clouds

	Clouds from 0 to 100 yards high.	From 100 to 200 yards high.	From 200 to 300 yards high.	From 300 to 400 yards high.	From 400 to 500 yards high.	From 500 to 600 yards high.	From 600 to 700 yards high.	From 700 to 800 yards high.	From 800 to 900 yards high.	From 900 to 1000 yards high.	From 1000 to 1050 yards high.	Above 1050 yards high.	Number of observations.
Jan.	0	9	12	28	53	39	37	32	30	39	36	116	431
Feb.	5	10	5	15	41	45	45	27	43	38	29	94	397
Mar.	2	1	6	11	22	40	32	36	24	32	44	184	434
Apr.	0	4	5	18	24	34	37	26	23	38	35	206	450
May	0	1	4	8	13	31	22	25	30	34	27	270	465
June	0	2	2	6	24	24	29	21	34	41	34	233	450
July	0	2	2	18	35	36	35	25	35	48	38	191	465
Aug.	0	4	5	13	27	39	35	26	25	45	30	215	464
Sep.	0	1	7	13	38	38	32	30	27	51	27	186	450
Oct.	2	0	5	13	26	49	31	31	46	61	37	164	465
Nov.	0	0	3	13	30	58	42	38	46	45	47	128	450
Dec.	1	8	6	23	41	53	39	50	47	46	35	111	460
Total	10	42	62	179	374	486	416	367	410	518	419	2098	5381

It may be proper to obſerve, that the ſuppoſition of the clouds riſing or falling with the barometer, or as the denſity of the air increaſes or diminiſhes, is not at all countenanced by theſe obſervations.—Alſo, that in very heavy and continued rains, the clouds are moſtly below the ſummit of the mountain; but it frequently rains when they are entirely above it.

G   SECTION

SECTION SIXTH

Account of Thunder-ftorms and Hail-fhowers.

WE fhall arrange the dates and accounts of
thefe, in the order of their fucceffion.
When the diftance of the thunder is mentioned,
it is calculated by obferving the number of fe-
conds between feeing the lightning and hearing
the thunder, and allowing 1142 feet of diftance
for every fecond of time.

Thunder-ftorms at Kendal *and* Kefwick.

1788.

May 26. Several loud peals of thunder a little before 7,
and again before 9. P. M. the laft very near, at Kendal.
The fame at Kefwick, at 7 P. M. with a few drops of rain.
—The ftorm from the SE.

July 3. From 6 to 7 P. M. much thunder, and very heavy
fhowers at both places It came from the S.

Auguft 15. From 7 to 8 P. M. thunder and heavy rain,
from the NW. at Kefwick.

Auguft 16 At $7\frac{1}{2}$ P. M. a tremendous ftorm paffed on
the SE. of Kendal, 8 or 10 miles diftant ; 20 or 30 flafhes
and reports fucceeded each other in about half an hour.

September 26. Diftant thunder in the night, at Kendal.
At $7\frac{1}{2}$ P. M. 2 claps at Kefwick, with much rain.

1789.

1789.

April 27. At 3½ P. M. fome loud peals of thunder, at Kendal.

May 13. From 6½ to 7 P. M. feveral loud claps of thunder, diftant, at Kendal.—Between 7 and 8 P. M. much thunder heard at Kefwick, from the SW.

May 17. A little before 3 P. M. one clap of thunder heard at Kendal.

June 12. Diftant thunder in the evening at Kendal.

—— 19. Diftant thunder P. M. at Kendal.

—— 20. At 1 P. M. feveral claps at Kendal;—the ftorm returned at 4 P. M. and there were 35 peals in ¾ of an hour, many of them uncommonly loud, and near; there was rain in the mean time, but not heavy.

N. B. A woman was killed by lightning, in a houfe at *Sedbergh*, about 11 miles from Kendal.

June 27. Diftant thunder in the evening, at Kendal.

July 4. Diftant thunder at 2 P. M. at Kendal.—Loud thunder, and heavy fhowers, P. M. at Kefwick.

July 6. After 2 P. M. diftant thunder, at Kendal.

—— 10. At 3 P. M. a diftant thunder clap, at Kendal.

—— 19. At 2½ P. M. diftant thunder, at Kendal.

—— 21. Paft 5 P. M. 3 loud peals, at Kendal; and diftant thunder, at Kefwick.

Auguft 29. P. M. fome thunder, with heavy rain, at Kefwick.

September 29. After 9 P. M. much diftant thunder, with fhowers, NW. at Kendal.—At 8½ P. M. one long and loud peal, at Kefwick.

September 30. Diftant thunder in the night, at Kendal.

1790.

April 26. At 1 P. M. fome peals of thunder, at Kendal.

May 16. At 9 P. M. one loud crack, from the E. at Kefwick.

1790.

May 17. At 11½ A. M. one loud crack, and a heavy fhower, at Kefwick.

June 9. From 6 to 10 P. M, much thunder, with little rain, at Kendal.

June 16. Diftant thunder in the evening, at Kendal*.— At 8½ P. M. feveral loud claps, at Kefwick.

June 22. From 6 to paft 8 A. M. much loud thunder, with rain, at both places.

Auguft 27. Some thunder P. M. at Kendal.

September 3. P. M. a little thunder, at Kefwick.

1791.

January 5. Loud thunder in the night, with hail, at Kefwick.

May 21. At 6 P. M. diftant thunder, and hail fhowers, at Kendal.

June 4. Betwixt 1 and 2 P. M. feveral peals of thunder, at Kendal. The laft of them was the moft remarkable one ever remembered at this place;—inftantaneoufly after the flafh, was heard a very loud and tremendous crack, exactly fimilar, but incomparably more loud, than the report of a mufket; every houfe in the town was fenfibly fhaken by it, and univerfal terror produced by the loudnefs and fingularity of the report; but providentially no harm was done.—The rain, mixed with hail, exceeded in quantity what has ever been produced here on a fimilar occafion, for 6 years at leaft; there fell upwards of *one inch and a half* in the fpace of *three quarters of an hour*, though a confiderable part of that interval was moderate rain.

N. B. It is remarkable that the barometer was ftationary all that day, and fo high as 30.06.

1791.

---

* There was, this evening, about Prefton-hall, 6 miles from Kendal, one of the moft extraordinary torrents of hail and rain, attended with thunder, that is upon record.

1791.

June 12. At 4 P. M. a crack of thunder, with hail and rain, at Kendal.

July 17. At 10 P. M. loud claps to the NW. at Kefwick.

——18. After 2 P. M. feveral claps, at Kendal; one of which not unlike that of the 4th of June. At Kefwick, 2 claps A. M. and 3 P. M. with exceffively heavy fhowers.

Auguft 15. Between 8 and 9 P. M. there was the moft lightning I ever remember to have feen at one time, at Kendal; fome thunder was heard, but it was diftant, E.

Auguft 16. From 5½ to paft 7 A. M. much thunder at both places, and heavy rain at Kefwick.

October 20. At 8½ A. M. one loud clap of thunder, at Kefwick, and much lightning from 7 to 10 P. M.—heavy rain all day.

December 25. Much thunder from 5 to 7 P. M. at Kefwick.

1792.

April 13. At 3 P. M. much diftant thunder, at Kendal.

May 27. Between 3 and 4 P. M. fome thunder and rain, at Kendal.

July 9. At 7 P. M. diftant thunder, at Kendal.

——16. P. M. much thunder, at Kendal. Between 6 and 8 P. M. loud thunder, at Kefwick.

July 18. At 8 A. M. thunder, at Kendal.

——25. After 6 P. M. thunder, at Kendal.

Auguft 26. At 3 P. M. fome thunder at Kendal.

October 14. In the evening, lightning; and at 10, diftant thunder, at Kendal. From 6 to 11 the fame evening, lightning, at Kefwick; and at the later hour, one long and loud crack of thunder,

Days

Days on which HAIL *has been noted in the journals
at* Kendal *and* Kefwick.

Hail at Kendal.	Hail at Kefwick.
1788. January 18.	1788. April 4. Nov. 4. Dec. 26 & 31.
1789. Jan. 18. Mar. 9. April 26 Oct. 1. Nov. 14. Dec. 15 & 16.	1789. Jan. 15. Feb. 11. April 1, 11, & 24. June 27. Sep. 14 & 30. Oct. 1 & 30. Nov. 13. Dec. 15, 16, & 31.
1790. Jan. 27. April 25. Aug. 3. Dec. 11 & 15.	1790. Feb. 16. April 11 & 14. July 31. Sep. 2 & 3. Dec. 11 & 13.
1791. Jan. 5, 7, 11 & 13. Feb. 1 & 11. Mar. 21. May 21, 22 & 23. June 4 & 12.	1791. Jan. 2, 4 & 5. Feb. 11, 15 & 18. Mar. 21. May 18, 20, 22, 23, & 25. June 12 & 21. July 5. Oct. 8 & 24 Nov. 5, 16 & 28.
1792. Mar. 19. May 22. Oct. 17. Dec. 6 & 22.	1792. Mar. 7. May 1 & 2. June 30. Sep. 20 & 21. Oct. 18. Nov. 15.

N. B. The winds that bring hail-fhowers are always SW,
W, or NW, in thefe places; and the barometer is generally
low.

In order to difcover what particular months or feafon of
the year, is moft liable to thunder-ftorms and hail-fhowers,
we have collected the feveral obfervations, at both places, in
each month of the year, into one fum, and placed them below.

	Jan.	Feb.	Mar.	Apr.	May	June	July	Aug.	Sep.	Oct.	Nov.	Dec.
Thunder	1	0	0	3	7	5	12	7	4	2	0	1
Hail	11	7	5	8	11	6	2	1	6	7	7	13

SECTION

SECTION SEVENTH.

Obſervations on the Winds.

I Have before obſerved, that my obſervations on the winds refer them all to 8 equidiſtant points of the compaſs, and to 5 degrees of ſtrength, marked 0, 1, 2, 3, and 4, reſpectively. Mr. *Croſthwaite* has referred them to 32, or the whole number of points, and to 12 degrees of ſtrength; but I have reduced his obſervations to agree with my own, in order to prepare the following table of compariſon.

The obſervations at both places were made three times each day, namely, morning, noon, and evening.

It may be obſerved, that the high winds do not in general differ materially, either in ſtrength or direction, at *Kendal* and *Keſwick*, as might be expected from the proximity of the places; but when the wind is moderate, there is often a difference in direction; probably the mountainous ſituations of the places may have ſome influence in this laſt caſe.

Here follows Tables containing the number of obſervations on the winds each year, in all the different directions, at both places.

WINDS

WINDS AT KENDAL.

Years.	N.	NE.	E.	SE.	S.	SW.	W.	NW.	Numb. of Obſervati.
1788	131	139	40	79	91	186	84	87	837
1789	94	118	38	49	94	309	76	46	824
1790	100	195	17	21	25	329	137	47	871
1791	62	259	16	33	19	440	138	50	1017
1792	51	294	33	24	35	472	92	33	1034
Total	438	1005	144	206	264	1736	527	263	4583

WINDS AT KESWICK.

Years.	N.	NE.	E.	SE.	S.	SW.	W.	NW.	Numb. of Obſervat.
1788	46	50	158	98	137	105	238	113	945
1789	53	47	150	120	180	146	211	119	1026
1790	32	62	143	105	134	174	237	89	976
1791	44	73	133	66	117	225	257	67	982
1792	49	84	139	88	164	213	219	49	1005
Total	224	316	723	477	732	863	1162	437	4934

To theſe tables we ſhall ſubjoin an account of thoſe days on which the higheſt winds prevailed, at one or both places.

Higheſt winds, marked 4, at Kendal *and* Keſwick.

1788.

Jan. 19. March 16. April 1 and 3. Dec. 26 and 27.

1789.

Jan. 13. Feb. 2, 3, 4, 11, 15, and 24. Oct. 1. Nov. 18. Dec. 15, 18, 19, 20, 24, 25, and 30.

1790.

Jan. 11. Feb. 12 and 26. March 10. June 19. July 5, 20, and 21. Oct. 12. Dec. 15 and 23.

1791.

1791.

Jan. 4, 5, 7, 8, 11, 12, 13, 15, 17, 18, 19, 24, 25, 29, and 30.—N. B. Theſe winds were all W. or SW. except on the 18th and 19th, SE. Feb. 1, 10, 11, 12, 15, 18, 19, and 22. March 4, 13, 19, 21, and 23. May 17 and 19. June 16. Oct. 20. Nov. 9, 11, 12, 19, 26, and 27. Dec. 1, 13, and 25.

1792.

Feb. 2. March 18. April 2, 15, 22, and 23. Sept. 10. Oct. 1, 2, 3, and 31. Nov. 18, 19, 20, and 21. Dec. 4, 5, 6, 9, 10, 11, 18, 20, 22, and 23.

In order to determine what months of the year are moſt liable to high winds, we have found the amount of the number of days in the ſeveral months, on which the higheſt winds were obſerved, according to the above account.

Jan.	Feb.	Mar.	Apr.	May	June	July	Aug.	Sep.	Oct.	Nov.	Dec.
18	17	8	6	2	2	3	0	1	7	12	24.

SECTION EIGHTH.

Account of the firſt and laſt appearance of Snow, each winter; the Froſt, Snow, ſeverity of the Cold, &c.

MOST people know that ſnow firſt appears in general upon the mountains; and the higher theſe are, all other circumſtances being the ſame, the ſooner their ſummits are covered.

H   with

with fnow; if they exceed a certain height (which varies with the latitude) fnow continues upon them all the year round, or is perpetual; but this is not the cafe with any mountains in *England.*

The higheft mountains feen from *Kendal* are to the NW. and do not exceed 6 or 7 hundred yards in height, as has been obferved; it is thefe of courfe that are firft topped with fnow. The mountains in the neighbourhood of *Kefwick* are much higher.

The firft appearance of *boar froft*, each autumn, has been pretty carefully noted, but the laft appearance of it, in the fpring, has not, it being inconvenient at that feafon to make obfervations previous to the rifing of the fun.

The dates of the different appearances follow for each year, together with the mean times; or, thofe times before or after which, upon an equality of chance, the events may be expected in future.

	Laft fnow feen on the mountains, in the fpring.		The fummits of the mountains covered with fnow.		The firft hoar froft on the grafs.	
	Kendal.	*Kefwick.*	*Kendal.*	*Kefwick.*	*Kendal.*	*Kefwick.*
1788	May 30	June 6	Nov. 15	Nov. 13	Sep. 15	Sep. 15
1789	May 14	June 30	Oct. 13	Oct. 29	Sep. 17	Sep. 17
1790	April 25	April 27	Nov 22	Oct. 31	Sep. 8	Sep. 4
1791	June 12	June 12	Oct. 22	Oct. 22	Oct. 13	Oct. 13
1792	May 1	Mar. 13	Nov. 15	Oct. 9	Sep. 16	Sep. 15
Mean	May 16	May 17	Nov. 8	Oct. 27	Sep. 20	Sep. 19

1788.

1788.

IN the beginning of this year there was very little froft or fnow; the moft fnow was on the 7th of March, being above 2 inches deep, both at *Kendal* and *Kefwick*.

In the beginning of December the froft fet in, and continued for 5 weeks; the mean ftate of the thermometer for which time was 28°; and at the end of it the froft had penetrated 16 or 18 inches into the ground.—Above 3 inches of fnow fell on the 31ft.

1789.

Not much fevere froft after the middle of January.—Snow on the 14th and 21ft of the fame. Much fnow from the 9th to the 14th of March; about 6 inches deep, at an average, both at *Kendal* and *Kefwick*.

Froft in November; very little in December.

1790.

Little either froft or fnow, in the beginning of the year.

On the 17th, 18th, and 19th of December, much fnow, 4 inches deep, at an average, at both places.

1791.

But little froft or fnow in the beginning of the year.

On the 8th, 9th, and 10th of December, a very great quantity of fnow; the average depth, at *Kendal*, was 11 inches, which was the greateft obferved there for 24 years paft; the average depth at *Kefwick* was about 8 inches.

It was on the evening of the fucceeding day, the 11th, that the extreme of cold took place; the air was clear, and the wind from the N. but very moderate; the barometer was 29.75; it was rifing before this event, and it fell afterwards. At *Kendal*, the thermometer at 8½ P. M. was — 6°, upon the fnow; afterwards it fell to — 10°; in the

morning

moʀning of that day it was 15°, and 20° at noon.—During the extreme cold, a prodigiouſly denſe miſt was carried from the river into the town, in which the thermometer fell no lower than 3°, whilſt it was — 10° to the N. of the river, and the air quite clear. The next morning the thermometer was at 18°, and the day windy, with ſhowers of ſnow, hail, and rain.

Probably the cold at *Keſwick* was as extreme as at *Kendal.* Mr. *Croſthwaite's* loweſt obſervation was 6°; but the proximity of his thermometer to the houſe, might be a means of keeping up the temperature in ſuch an extremity as this.

1792.

Strong froſt the ſecond week in January.
Little froſt or ſnow in November and December.

═══════════

SECTION TENTH.

Account of Bottom winds *on Derwent lake.*

DERWENT lake is one of thoſe few which are agitated at certain times, during a calm ſeaſon, by ſome unknown cauſe. The phenomenon is called a *bottom wind.*

Mr.

Mr. *Crofthwaite* has been pretty affiduous in procuring intelligence refpecting thefe phenomena, and in obferving any circumftance that might lead to a difcovery of their caufe; but nothing has occurred yet that promifes to throw light on the fubject.

N. B. The *lake* is near *Kefwick.*

The following is an account of the times and circumftances of the feveral obfervations.

1789.

April 30. From 8 A. M. till noon, the lake pretty much agitated.

Auguft 9. At 8 A. M. the lake in very great agitation; white breakers upon large waves, &c. without wind.

Auguft 27. At 9 A. M. a fmall bottom wind.

1790.

June 20. At 8 P. M. a bottom wind on the lake.

October 11. At 8 P. M. a bottom wind on the lake.

December 1. At 9 A. M. a ftrong bottom wind on the lake.

1791.

The phenomena that took place this year, if any, were not noticed.

1792.

October 28. At 1 P. M. a bottom wind; the water much agitated.

SECTION

SECTION ELEVENTH.

Account of the Auroræ Boreales *seen at* Kendal *and* Kefwick.

THE *aurora borealis*, or that phenomenon which in *England* is called the *Northern lights*, or *ftreamers*, has appeared frequently to all the northern parts of *Europe* fince the year 1716, though it feems to have been a rare phenomenon before that time.

Sometimes the appearance is that of a large, ftill, luminous arch, or zone, refting upon the northern horizon, with a fog at the bottom; at other times, flafhes, or corufcations, are feen over a great part of the hemifphere.—We fhall defcribe the general phenomena more at large in the effay on the fubjeƈt, in the fecond part of this book; and particular obfervations will be given at large in the addenda to this feƈtion.

Explanation of the following Lift.

IN the firft column we have given the month and day on which the *aurora* was feen; in the fecond, the hour P. M.; when no hour is mentioned, it is to be underftood to have happened between the end of the twilight and 10 o'clock. The third column contains the moon's age at the time, or the number of complete days betwixt the *change*

and

and the *aurora*; the fourth contains the days in like manner betwixt the *full* and the *aurora*; the reasons for these two columns will appear in the Essay. In the fifth column we have characterized the *aurora*, by one or more words: *still*, denotes the northern horizontal arch; and *active*, denotes those appearances when distinct flashes and coruscations were seen. but this distinction was not always attended to, and if it had, the *aurora* often exhibits both appearances at the same time; *grand*, denotes a large display of streamers over great part of the hemisphere; *high*, denotes near the zenith, and *low*, near the horizon, apparently.

N. B. The dates of those observations not characterized, I received from a friend; they may be depended on as authentic.

A List of the Auroræ Boreales *observed at* Kendal *and* Keswick, *for 7 years, namely from May 1786 to May 1793, together with the moon's age at the respective times of observation.*

N. B. For distinction's sake we have marked all those that were observed at both places with 2, and those observed at *Keswick* only, with 1; the rest were observed at *Kendal* only.—Those marked D, were doubtful observations, from twilight, or other causes.

1786.	Hour P.M.	☽'s age.	☽ past full.	Character.	1786.	Hour P.M.	☽'s age.	☽ past full.	Character.
May 1		3			Sep. 21		29	14	
——11		13			——26		4		
——22		24	9		——29		7		
July 15		20	4		Oct. 13		21	6	
Aug 11		17	2		——25		3		
——17		23	8		Nov. 14		23	8	
Sep. 8		16	1		Dec. 25		5		active, low.
——19		27	12						
——20		28	13					Number 16.	

1787.	Hour P.M.	D's age.	D past full.	Character.	1788.	Hour P.M.	D's age.	D past full.	Character.
Jan. 12		23	9		Jan. 13	8	5		tranfient
—24		5			—14	8	6		large, ftill
—25		6			—15	9	7		large, ftill
Feb. 22		4			Feb. 4	9	27	12	ftill
Mar. 21	8	2		active, high	— 6		29	14	active, high 1
—24	8	5		active, high	— 7		0		fmall 1
Apr. 19	9	1		high. D	— 8		1		faint, ftill
—20		2		high. D	—12		5		active, fmall 1
—26		8		active, low	Mar. 7	11	0		active, fmall 2
May 12		24	10	faint. D	— 8		1		active, fmall
—16	9	28	14	high. D	—28	10	21	6	bright, large
—17	9	0		active	Apr. 1	10	25	10	
—18	11	1		active	— 3	10	27	12	large, grand
June 7	11	21	7	active	— 7	10	1		a glance, clouds
Aug. 7	9	24	9	active	—14	12	8		
—19	10	6		active	—27	10	21	7	ftill, low
Sep. 19	9	8		bright, ftill	—28	10	22	8	active, high 2
Oct. 4	10	22	7	ftill	—29	10	23	9	large, active
— 6	10	24	9	active, faint	—30	10	24	10	ftill
— 7	10	25	10	active, large	May 1	10	25	11	a glance, clouds
—17	11	6		active (a)	— 4	10	28	14	tranfient
—19	9	8		ftill	—10	10	4		high. D
Nov. 4	9	24	9	large, bright	—11	10	5		large, ftill
— 8	8	28	13	large, bright (b)	—24	10	18	4	very grand (c) 2
—28	9	19	3	ftill, fmall	—25	11	19	5	grand
—29	9	20	4	ftill, fmall	—27	10	21	7	active
—30	9	21	5	ftill, fmall	June 3		29	14	large. D
				Number 27.	July 30		27	12	active
1788.					Aug. 1	10	0		active (d)
Jan. 9	6	1		ftill, low	— 2	10	1		active
—10	8	2		ftill, large 2	— 3	10	2		fmall
—11	8	3		ftill, faint	—19	10	18	3	large, ftill 2

1788.

(*a*) A heavy fhower, with thunder, juft before.

(*b*) Several flafhes of lightning with it, after a very wet day.

(*c*) From 10 to 11 P. M. uncommonly brilliant, active ftreamers over moft of the hemifphere : they were faid to be heard.----Not much inferior the next night.

(*d*) Splendid ftreamers, extent from NE. to W. ; no fog beneath.

1788.	Hour P.M.	D's age.	D paft full.	Character.	1789.	Hour P.M.	D's age.	D paft full.	Character.
Aug 23		22	7	very grand (a) 2	—29		3		a glance, clouds
—29	10	28	13	active	—30		4		a glance, clouds
Sep. 2		2	2	ftill	Apr. 12		17	2	ftill
—6		6	6	a glance, clouds	—13		18	3	ftill 2
—10		10	10	faint, ftill	—30		5		ftill
Oct. 12	4 M	12		ftill	June 12	10	19	5	active
—21	10	22	6	ftill	Aug 13		22	8	active, high 1
—24		25	9	ftill	—14	10	23	9	active, high
—27		28	12	ftill	—15	9	24	10	ftill
—30	9	1		ftill	—16	10	25	11	ftill
—31	9	2		large, ftill	—17	10	26	12	ftill
Nov. 1		3		ftill	—18	10	27	13	active
—19	9	21	6	ftill	—19	10	28	14	active, high
—27		0		faint, ftill	—20	10	0		active
—28		1		bright, ftill	—25		5		active 1
—30	7	3		ftill	Sep. 14	10	24	10	ftill (d)
Dec. 21		24	8	ftill 2	—15	10	25	11	ftill
—24		27	11	active	—20	10		1	fine, active 2
				Number 53	—23		4		fine, clouds
1789.					—26		7		grand (e) 2
Jan. 11		15	0	ftill	Oct. 18		0		active
Feb 15		20	5	active, fmall 1	—19	11	1		active
—23	11	28	13	active	—20		2		grand (f)
—26	10	1		large, active	—23		5		large, ftill 2
—28		3		large, bright (b)	—25		7		ftill
Mar. 14		17	3	very grand (c) 2	—27		9		ftill
—16		19	5	ftill, fmall	—31	8	13		ftill

I　　　　　　　　　　　　　　　　　　1789.

(a) Soon after 8 P. M. a broad arch was obferved, extending quite acrofs the heavens, through the zenith, from E. to W. nearly ; but its eaftern extremity inclined to the north, and its weftern to the fouth: afterwards an uncommonly grand difplay of ftreamers over two-thirds of the hemifphere.

(b) At 9½ P. M. there was a large bow, like that of the 23d of Auguft laft.

(c) It began S. of the prime vertical, and afterwards fpread northward.

(d) Flafhes of lightning, both this evening and the fucceeding.

(e) Moft of the hemifphere finely illuminated with ftreamers.

(f) From 8 to 10 a grand difplay of ftreamers over great part of the hemifphere.

1789.	Hour P.M.	☽'s age	☽ paſt full	Character.
Nov. 4		17	1	ſtill
—10	11	23	7	ſtill
—14		27	11	bright(*a*) 2
—19		2		large, ſtill 2
—21		4		active 2
—22		5		fine, active.
—24	11	7		ſtill
—25	10	8		ſtill
—26		9		ſtill
—27	10	10		active
Dec.14		27	12	ſtill, clouds
				Number 45
1790.				
Jan. 14	7	29	13	ſtill
Feb. 3		19	4	ſtill
—9		25	10	ſtill
Mar.10		24	9	ſtill, faint
—16		1		ſtill
—17		2		ſtill, faint
—18		3		ſtill
—19		4		ſtill, bright
—20		5		ſtill, bright
Apr. 3		19	4	ſtill, low
—4		20	5	active, low
—5		21	6	ſtill, faint
—6		22	7	ſtill
—7	11	23	8	ſtill
—9		25	10	ſtill
—16	11	2		active
—17	10	3		large, ſtill
May 12		28	13	active
—14		0		ſtill
—16		2		ſtill
—18		4		ſtill
Sep. 7	9½	28	14	ſtill
Oct. 9		1		ſtill, faint
—18	7½	10		ſtill 1

1790.	Hour P.M.	☽'s age	☽ paſt full	Character.
Oct. 31	8	23	8	ſtill
Nov. 7	10	1		ſtill
—8	9	2		ſtill
—9		3		full
—10	10	4		ſtill
—12	10	6		ſtill
—16	10	10		ſtill
—27	8	21	6	fine, active 2
—28	8	22	7	ſtill, faint
—30	10	24	9	ſtill, faint
Dec. 25		19	4	ſtill, faint
—28		22	7	ſtill, faint
				Number 36
1791				
Jun 6	9	2		very grand 2
Feb. 25		22	7	bright, ſtill
Mar. 3		28	13	ſtill
—5		1		ſtill
—7		3		ſtill 2
—26		22	6	large, ſtill
—29		25	9	ſtill, low
Apr. 3		0		ſtill, ſmall
—20		17	2	ſtill
—23		20	5	ſtill
—25		22	7	ſtill, ſmall
May 12	10	9		active
—20	11	17	2	active
June 10		9		active
Sep 5		7		active
—8	10	10		ſtill, ſmall
—11	9	13		ſtill
—13		15	1	ſtill
—27		0		ſtill
—28		1		ſtill
Oct. 15		18	3	ſtill
—19		22	7	ſtill
—20		23	8	active (*b*)

1791.

(*a*) Lightning afterwards.   (*b*) Thunder at a diſtance this evening.

1791	Hour P.M.	☽'s age.	☽ paſt full.	Character.	1792	Hour P.M.	☽'s age.	☽ paſt full.	Character.
Oct. 22		25	10	ſtill, bright 2	June 30	1 M	10		active
—23		26	11	ſtill	Aug. 4	10	16		2 active, ſmall
—29		2		large, bright	—23	10	6		active
—31		4		active	S-p. 22	10	6		bright ſtill
Nov 3		7		large, active 2	Oct 12	10	26		12 ſmall, ſtill
— 4		8		ſtill, faint	—* 13		27		13 very grand 2
— 5		9		ſtill, faint	—14		28		14 active 2
—11		15	1	ſtill	—18	10	3		ſmall
—14		18	4	ſtill	—23		8		ſtill
—17		21	7	ſtill	—31	8	16		2 active, low
—18		22	8	ſtill	Nov. 19	9	4		ſtill 2
Dec. 13	6₂	18	3	fine, active	Dec. 7		23		ſtill 1
—19		24	9	ſtill					Number 23.
—26			1	ſtill	1793.				
				Number 37.	Jan. 11	10	29	14	ſtill, ſmall
1792.					—12		0		active
Jan. 9	10	15	0	large, ſtill	—13		1		ſtill
—17		23	8	ſtill, faint	Feb. 8		27	12	ſtill
—18		24	9	ſtill, faint	—12		2		grand 2
Feb. 9		17	1	ſtill	—15		5		an arch 2
—17		25	9	ſtill	Mar. 5		23	8	ſtill
Mar. 2		9		ſtill	— 6		24	9	ſtill
—15		22	7	large, bright	—13		1		ſtill 2
Apr. 10		19	3	very grand 2	—30		18	3	fine, high 2
—11		20	4	grand 2	Apr. 5		24	9	ſtill
—16		25	9	ſtill. bright	— 9		28	13	active 2
May 6		15	0	active	—14	12	5		ſtill 1

General obſervations on the Auroræ before October 13, 1792.

IN making obſervations upon any phenomenon in nature, with a view to aſcertain its cauſe, every particular circumſtance ſhould be attended to ; for, though many may be

I 2 found

* A more particular account of the ſucceeding ones will be given hereafter.

found afterwards to be trivial, and of little or no moment in
leading us towards the difcovery, yet fome one or other of
them generally happens to be of importance. It will be
feen hereafter, that the exact bearing and extent of the large,
ftill, horizontal arch of the *aurora*, and the point in the heavens to which the corufcations tend, are amongft the cir-
cumftan es of much importance in the inveftigation of its
caufe. Thefe circumftances, it muft be confeffed, were not
accurately noticed, either at *Kendal* or *Kefwick*, previous to
the middle of October, 1792.

As for myfelf, the only minute I ufually made upon the
ftill aurora was, that it was fituate in the NW. by which I
meant that its centre was between the N. and the W. without once attempting to afcertain the exact bearing of the
centre ; and the *corona*, when there was one, is often mentioned in my notes, as being fouth of the zenith, but the
number of degrees was not afcertained.

Mr. *Crofthwaite*, however, has been rather more particular
at times with refpect to the bearings, extent, &c. The
centre of that on January 10, 1788, he obferves bore NNW.;
that of the 28th of April, NW. b N. ; the centres of all
the reft are faid to have been between the North and Weft,
or elfe North ; not one was obferved to have its centre to
the Eaft of the meridian.

*N. B. The additional obfervations on the Auroræ, beginning
with that on the 13th of October, 1792, will be given after the
next Section.*

SECTION

SECTION TWELFTH.

On Magnetifm, and the variation of the Needle.

IN order to underſtand the additional obſervations, and the ſubſequent Eſſay on the
aurora borealis, a competent knowledge of magnetiſm is requiſite ; and as the principal facts
relating to that ſubject are few and ſimple, we
have thought it would not be amiſs to ſtate
them here, for the ſake of ſuch as may not be
previouſly acquainted therewith.

The *Loadſtone*, or *natural Magnet*, is a mi-
neral production, found in the bowels of the
earth, amongſt rich iron ores, of which it is one
itſelf ; its diſtinguiſhing property is that of attracting iron and ſteel. This property, which is
called magnetiſm, is communicable to *ſteel* only,
ſo as to be permanent ; and to *iron* when within
the influence of a magnet, but as ſoon as the
magnet is withdrawn, the magnetiſm of iron
ceaſes.

Every magnet has two oppoſite points or extremities, called its *poles ;* the one is denominated its *north pole*, and the other its *ſouth pole ;*
and the attraction of the magnet is ſtrongeſt at
its poles.

If

If an oblong bar of tempered fteel (it will anfwer well if 5 inches long, half an inch broad, and a quarter of an inch thick) be rubbed over from one end to the other, always the fame way, by either pole of a magnet, it will be converted into a magnet itfelf; and that end to which the pole was firft applied, will be a pole of the new magnet, of the fame name as the generating pole. By rubbing the new magnet the contrary way, with the fame pole, its magnetifm will be firft deftroyed, and then frefh magnetifm will be communicated; but the poles of the new magnet will be of contrary names to what they were before.

Either pole of a magnet attraɛts iron, or fteel not magnetic; but the pole of one magnet, *re-pels* the pole of another magnet, of the fame name, and *attraɛts* the pole of a contrary name; the repulfion in the former cafe feems to be equal to the attraɛtion in the latter.

Magnetifm is fometimes communicated, de-ftroyed, or inverted, by lightning, or by an eleɛtric fhock, &c.

If a magnetic bar, or needle, be fuffered to move freely in an horizontal plane, it will only reft in one pofition, when the north pole points northward, and the fouth pole fouthward — Hence the common needle and compafs, which

was

was invented about the beginning of the 14th century.

If a plane perpendicular to the horizon be conceived to be drawn through the horizontal needle, when at reft, it is called the plane of the magnetic meridian; and the angle made by this plane, with the plane of the true meridian, is called the *variation of the needle.*

If a magnetic needle be nicely poifed on an axis paffing through its centre of gravity, or middle, and fuffered to move freely both horizontally and perpendicularly, it will reft only in one pofition, namely, when in the plane of the magnetic meridian, and having its north pole pointing towards the ground; the angle of deflection from the horizontal plane, is called the *dip* of the needle, and the needle itfelf in this cafe a *dipping-needle;* its pofition is the proper and natural one of every magnet that is fuffered to be guided folely by the magnetic influence. From this phenomenon, and others of the fame nature, it is inferred, that the earth itfelf is a magnet; whether its magnetifm refults from the united influences of the natural magnets it contains, or whether its magnetifm may be in its atmofphere, is not certain; and as poles of unlike denominations attract each other, the fouth pole of the earth's magnetifm muft be in the northern hemifphere, becaufe it attracts the north pole of the needle.

The

The variation of the needle is very different at different places of the globe, and even at the fame place at different times; in thefe parts it is at prefent wefterly, and is increafing every year; the variation at *London* in 1580 was 11° 15′ E. in 1657 it was 0° 0′; at prefent, 1793, it is about 22°¼ W. and increafes nearly 10′ each year. From the refult of feveral obfervations I find it to be 25° W. at this time, at *Kendal.*

The dip of the needle too is very different at different places, and probably at the fame place at different times; but, for various reafons, the obfervations on this head are neither fo numerous nor fo accurate as thofe of the variation. It feems at prefent to be about 72° at *London,* according to Mr. *Cavallo;* and there is reafon to fuppofe, it is not many degrees different in any part of *England;* for want of proper inftruments I have not been able to afcertain it at this place.

Befides the annual change in the variation of the needle, there is a daily change, or variation of the variations. According to Mr. *Canton,* who made a feries of obfervations on the daily variation for a long time, the north pole of the needle moves gradually weftward till 2 or 3 P. M. and then returns gradually to its former ftation; the mean daily variation in winter is about 7′, and in fummer about 13′¼. He moreover obferved, that the needle was difturbed when an *Aurora borealis* was in the atmofphere.

I have

I have myself made a like feries of obfervations for fome months, and find them in general
to agree with his; but as it is not neceffary for
my purpofe to relate the refult of them, any further than what is contained in the fubfequent
pages, I fhall not detain the reader longer on
the fubject.

―――――

Addenda *to the Obfervations on the Auroræ Boreales.*

1792.

OCTOBER 13. At *Kendal*, A. M. frequent gleams.
P. M. hazy; from 4 till 8 rainy, at which time the clouds
to the *fouth* were remarkably red, and afforded fufficient
light to read with, though there was no moon, nor light in
the north. The unufual appearance raifed my curiofity,
and I waited with impatience to fee the clouds carried off
to the SE. (for the wind was W. or NW. and pretty frefh).
In the mean time, having by me a very good *theodolite*, made
by *Dollond*, I took it out to make obfervations on the bearing, altitude, &c. of any remarkable appearance.

From $9\frac{1}{2}$ to 10 P. M. there was a large, luminous, horizontal arch to the fouthward, almoft exactly like thofe we
fee in the north; and there was one or more faint, concentric arches northward,—It was particularly noticed, that
all the arches feemed exactly bifected by the plane of
the magnetic meridian. At half paft 10 o'clock, ftreamers
appeared very low in the SE. running to and fro from W.
to E, they increafed in number, and began to approach the

K   zenith,

zenith, apparently with an accelerated velocity; when, all
on a fudden, the whole hemifphere was covered with them,
and exhibited fuch an appearance as furpaffes all defcription.
—The intenfity of the light, the prodigious number and
volatility of the beams, the grand intermixture of all the
prifmatic colours in their utmoft fplendor, variegating the
glowing canopy with the moft luxuriant and enchanting
fcenery, afforded an awful, but at the fame time, the moft
pleafing and fublime fpectacle in nature. Every body gazed
with aftonifhment; but the uncommon grandeur of the fcene
only lafted about one minute; the variety of colours difappeared, and the beams loft their lateral motion, and were
converted, as ufual, into the flafhing radiations; but even
then it furpaffed all other appearances of the *aurora*, in that
the *whole* hemifphere was covered with it.

Notwithftanding the fuddennefs of the effulgence at the
breaking out of the *aurora*, there was a remarkable regularity obfervable in the manner.—Apparently a ball of fire ran
along from E. to W. and the contrary, with a velocity fo great
as to be but barely diftinguifhable from one continued train,
which kindled up the feveral rows of beams one after another; thefe rows were fituate one before another with the
exacteft order, fo that the bafes of each row formed a circle
croffing the magnetic meridian at right angles; and the feveral circles rofe one above another in fuch fort that thofe
near the zenith appeared more diftant from each other than
thofe towards the horizon, a certain indication that the real
diftances of the rows were either nearly or exactly the fame.
And it was further obfervable, that during the rapid lateral
motion of the beams, their direction in every two neareft
rows was alternate, fo that whilft the motion in one row was
from E. to W. that in the next row was from W. to E.

The point to which all the beams and flafhes of light uniformly tended, was in the magnetic meridian, and, as near
as could be determined, between 15 and 20° fouth of the
zenith.—The *aurora* continued, though diminifhing in fplendor, for feveral hours. There were feveral meteors (falling
                                                    ftars)

ſtars) ſeen at the time; they ſeemed below the *aurora,* and
unconnected therewith.———It was ſeen at *Keſwick, Leeds,*
&c. with much the ſame circumſtances; but how far it extended I have not learned.

The variation of the needle during the *aurora,* was not
noticed.

October 14. I did not notice the *aurora* myſelf this evening; there was thunder and lightning, both here and at
Keſwick, at the time of the *aurora.*

October 18. At Kendal. The *aurora* this night was an
oblong, luminous cloud, about 15 or 20° long, and 4 or 5°
broad, bearing about SE. by E. and 10 or 20° above the
horizon; its ſouthern extremity was higher than its northern,
and it evidently lay in the tract of a great circle from E. to
W.—It diſappeared ſeveral times, and reappeared again almoſt inſtantly; and ſeveral times waxed and waned without
vaniſhing; no radiations ſhot from it.

October 23. At Kendal. The *aurora* this evening appeared as an arch in the north-weſt quarter, from which
proceeded ſeveral beams; they converged to a point on the
magnetic meridian, about 18° beyond the zenith.

October 31. At Kendal. A few beams were ſeen to run
to and fro from E. to W. low, or near the horizon: the
moon ſhone bright at the time, and the clouds coming on
ſoon after, the whole was obſcured.

November 19. At Kendal, the particulars of the obſervation were miſlaid; at Keſwick, the *aurora* roſe to about
18° above the horizon, and was ſituate in the uſual quarter.

December 7. At Keſwick, a faint appearance; about
5° high.

1793.

January 11. At Kendal, a ſmall arch in the horizon; it
roſe to 5 or 10° altitude, and was biſected by the magnetic
meridian.

January 12. At Kendal, from 6 to 9 P. M. a horizontal,
luminous arch, 20° altitude, and biſected by the magnetic

K 2   meridian.

meridian. After 9, fine ftreamers ftruck out, and ran to and fro a while acrofs the faid meridian, and then were converted into flafhes, as ufual; fome rofe up to the zenith. The point of convergency, and every other particular, were, to all appearances, the fame as have been defcribed before.

The needle was confiderably agitated at the time.

January 13. At Kendal, very bright in the northern horizon, but clouded above.—The variation of the needle at 6 P. M. 25° W.; at 9 P. M. 24° 34′; at 10 P. M. 24° 54′; next morning 25° 4′.

February 8. At Kendal, bright northward at 8½ P. M. at 10, the luminous arch was 16° altitude.—The other circumftances relating to it follow, fuppofing the variation of the needle at the noon of that day 25° W.

H. M.	Variation of the needle.	
10 —— P. M.	25° 0′ W.	the arch rifing.
10 10 ——	24 54 —	bright ftreamers, low, with clouds.
10 30 ——	24 42 —	ftreamers rifen; fine, weftward*.
10 35 ——	24 37 —	a ftill light; clouded above.
10 45 ——	24 57 —	bright, eaftward; clouds above.
10 55 ——	25 7 —	light equal, eaft and weft.
11 5 ——	25 7 —	bright, low; clouded above.
11 15 ——	24 57 —	clouded, but bright eaftward.

It was related to the magnetic meridian as the former ones.

February 12. At Kendal, the *aurora* appeared foon after 6 P. M. flaming over two thirds of the hemifphere. The beams all converged to a point in the magnetic meridian, about 15 or 20° to the fouth of the zenith, as was found from frequent trials.—The other particulars follow.

H. M.	Variation.	
5 —— P. M.	25° 5′ W.	
6 35 ——	24 49 —	altitude of the clear fpace fouth 35°.
6 42 ——	24 55 —	alt. of do.-20°; ftreamers bright, eaft.
6 50 ——	25 — —	}
7 2 ——	25 28 —	} ftreamers bright and active all over
7 5 ——	25 12 —	} the illuminated part.

H.

———

* That is, relative to the magnetic meridian, here and elfewhere.

H. M.   Variation.

7 10 P. M. 24°40′ W. difappeared in the weft; active, eaft.

7 15 —— 24 40 —

7 20 —— 24 35 — active about the zenith; light faint.

7 25 —— 24 45 — light faint.

7 35 —— 24 45 — light faint.

8 —— —— 24 45 — ftrong light northward.

8 10 —— 24 45 —⎫ a large, uniform, ftill light, cover-
8 35 —— 24 47 —⎬   ing half the hemifphere, with
                    ⎭ flafhes now and then.

9 15 —— 24 43 — ftreamers NW. bright, eaft; clouds.

9 20 —— 24 43 — the *aurora* burfting out afrefh.

9 30 —— 24 50 —⎫ as fine and large a difplay of ftream-
10 —— —— 24 55 —⎭   ers as has appeared this evening.

10 15 —— 24 57 —⎫ the light growing fainter and
10 35 —— 24 40 —⎭   fainter.

8 — A. M. 24 57

N. B. The arch bounding the *aurora* to the fouth, was always at right angles to the magnetic meridian, when perfect.

At Kefwick, the fame evening; 7 P. M. ftreamers from ENE. to WSW. and 28° paft the zenith; perpendicular beam bore N. 17° W.—At 9h 25m very fine; they converged to a point 15° fouth of the zenith, bearing SSE.— Altitude of clear fpace 30°. The perpendicular beam N. 35° W.; extent on the horizon from ENE to WSW.— At 10h 30m they were fettled in the northern quarter into an arch of 13° altitude, whence ftreamers fhot up towards the zenith.

February 15. The *aurora* of this evening was feen both at Kendal and Kefwick, and, as far as the eye could judge, the appearances feem to have been the fame at both places. —It was a luminous arch, the centre of which bore SSE.; and it was extended in the oppofite directions of ENE. and WSW.: on the weft fide its extremity feemed to touch the mountains at both places, at the altitude of 6°; and on the eaft fide it extended about half way to the horizon. The eaftern end was rather ovaliform, about 8 or 10° broad, and

where

where it joined to the rest, was narrowest of all, being but
2 or 3° broad, and bearing SE. ; after which its breadth
increased towards the west, being in some places 6 or 8°.
—The sky was clear, and there was no appearance of an
aurora in the north, except two or three small streamers at one
time, quite in the horizon. The eastern end of the arch
waxed and waned frequently, and sometimes entirely vanished, and then reappeared again, in the space of a few seconds.
About a quarter past 10 it grew faint, and finally disappeared.
It did not sensibly vary in position during its appearance,
and just before it vanished, its situation amongst the stars, as
seen from Kendal, was as follows :—the south edge of the
arch seemed to touch pretty exactly the star *lucida colli*, or
gamma Leonis, to pass 4 or 5° above *Procyon*, thence through
the middle of the constellation *Orion*, leaving his bright foot,
Rigel, 2 or 3° to the south.

From these observations it results, that the greatest altitude of the edge, at Kendal, must have been about 53°.
Mr. *Crosthwaite* found the greatest altitude of the said edge,
at Keswick, to be 48°. The distance of the two places, as
has been observed, is about 22 English miles, and it fortu-
nately happens, that they lie very nearly in the direction of
a plane at right angles to the arch ; hence, we have the re-
quisite *data* to determine the height of the arch, which, by
trigonometry, comes out 150 English miles.

The parallactic angle being so small, an error of 1 or 2°
in the altitudes, is of great consequence.—Mr. *Crosthwaite*
thinks the error in his observations could not exceed 1°$\frac{1}{2}$,
as the light was steady at the place where the altitude was
taken.—Admitting the errors amounted to 2° at each place,
which exceeds the bounds of probability, and that they were
contrary ; we shall then find the height 83 miles in the one
case, and 750 in the other, which may, I think, be safely
considered as boundaries, betwixt which the true height
was ; and hence it may be inferred, that the arch would be
visible to all *Great-Britain* and *Ireland;* that it is much to be
wished, some persons in more distant places, may have made
similar

fimilar obfervations upon the phenomenon, by which it**s** height may be determined with more precifion.—In the mean time we fhall confider it as 150 Englifh miles.

March 5. The *aurora* at Kendal was feen at 8 P. M. ; it was a bright ftill light a while, but foon clouded.——The needle was not attended to.

March 6. At Kendal, a few fine ftreamers at 9 P. M. altitude 15°, and extent along the horizon 70° ; exactly bi-fected by the magnetic meridian. It foon dwindled into a faint light. At 9h 35m brighteft on the northern fide.—— The needle was 25° at 9h 4m,—24° 58′ at 9h 14m,—24′ 50′ at 9h 35m,—24° 55′ at 10h 30m,—24° 52′ at 8 the next morning.

March 13. At Kendal the needle was at 8h 30m 24° 30′, —at 10h 30m 25° 4′,—and at 8 next morning 25° 4′.—— There was a brightnefs northward at 10 P. M. but pretty much clouded ; this circumftance, with that of the needle, rendered it probable an *aurora* was in the atmofphere.—— It was confirmed by the following account.

At Kefwick, the fame evening, at 8h 18m a horizontal arch, extent from NW. by W. to NNE. with faint ftream-ers ; the arch 20°, and ftreamers 25° altitude ; the vertical ftreamers bore NNW. At 10, an arch from WSW. to ENE. its greateft altitude 30° : no ftreamers.

March 30. At Kendal, at 8 P. M. there appeared fome faint concentric arches of an *aurora ;* it was not further no-ticed till,

H. M.   Variation.

8 35 P. M. 25° 5′ W. a grand horizontal arch, altitude 6°.

8 40 —— 25 25 — ftreamers to 30° paft the zenith.

8 48 —— 25 5 — bright eaftward.

8 55 —— 25 5 — ftreamers faint.

9 — —— 25 5 — denfe light north ; rare above.

9 5 —— 25 10 — ditto

9 10 —— 24 55 — bright weftward.

9 15 —— 24 55 — a fine, perfect, horizontal arch.

9 20 —— 24 55 — altitude of its upper edge 30°.

H.

H. M.	Variation.	

9 30 P. M. 24° 30´ W. ſtreamers up to the zenith.
9 35 ——— 24 35 — diſperſed, and not ſo high.
9 45 ——— 24 58 — faint light; brighteſt eaſtward.
9 52 ——— 24 45 — dull light.
10 ——— 24 42 — dull light; haze below.
10 10 ——— 24 42 — haze riſen; light fainter.
10 15 ——— ——— — clouds riſen; light almoſt vaniſhed.
10 30 ——— 24 45 — clouds more riſen.
11 15 ——— 24 45 — ſeveral ſmall clouds cover the hemiſ-
8 —A. M. 24 54   [phere.

There were ſeveral fine, perfect, concentric arches north-ward, during moſt of the time.—At 8h 48m one fine arch, the altitude of its under edge 10°. At 8h 55m two perfect arches, altitude of the higher 12°, with a fine edge. At 9h ſeveral concentric arches, one with a fine edge, altitude 11°. At 9h 5m one of the upper arches with a very bright edge, its altitude 13°; the baſes of the ſtreamers compoſing it of very denſe light, and rare above. At 9h 10m its alti-tude 13 or 14°.—At 9h 15m the upper edge of the large horizontal light ſeems now as well defined as that of a rain-bow, its altitude 47°, and that of the under edge 10°. At 9h 20m altitude of upper edge 30°.

The arches were all at right angles to the magnetic meri-dian, and the beams had their uſual convergency.—At one time ſeveral ſmall ſtreamers formed a *corona* upon the mag-netic meridian, the centre of which was determined by a good obſervation to be 72° from the ſouth.

The ſky was free from clouds till the laſt.

At Keſwick, the ſame evening, at 8h 20m there were bright ſtreamers WNW.—At 8h 28m they had ſpread from WSW. to ENE.; altitude of the arch 14°; vertical ſtream-ers bore NW. by N. At 8h 35m ſtreamers 43° paſt the ze-nith*: previous to this there were at one time three con-centric

* By the obſervations at *Kendal*, the *aurora* was 30° paſt the zenith at 2h 40m, and the clocks being corrected at both places, ſo as to be near
the

centric arches northward, fet with bright ftreamers, which had a very quick lateral motion; the under edge of the higheft was not more than 14° high. At 9h 6m the altitude of the faid arch was 13°½, bearing NW. ¼ N.; ftreamers fhort, being only 5° higher than the under edge; horizontal extent of the arch from W. by N. to NE.

April 5. At Kendal, a fmall blufhing of light, exactly in the magnetic north, at 9 P. M.; it foon faded away.

The difturbance of the needle was imperceptible.

April 9. The *aurora* was firft feen at Kendal, at 9h 30m P. M. being a fmall blufhing of light in the magnetic north. At 10h the arch rifen to 6 or 8° of altitude, with ftreamers from 3 or 4 to 10° altitude, and a mift below; the reft of the fky was extremely clear; the light was denfe at the under edge of the arch. At 10h 25m bright and active weftward; the mift below.—Soon after, uncommonly active ftreamers, very low; the light feen denfe through the mift. At 10h 35m the mift vanifhed; the *aurora* rather larger, and duller. At 11h a larger arch, altitude 10°, with mift below; no ftreamers, the light being ftill and uniform. At 11h 10m ftreamers very active; their progrefs feemed down, or northward.

The needle was not fenfibly difturbed all the while.

At Kefwick, the fame evening, a faint light at 9h 45m. —It was 7° high at 9h 54m, and the higheft part bore N. by W. ¼ W.; one minute after, bright ftreamers from NE. to WNW. the greateft altitude of their bafe 5°⅓, the bearing of the fame NW. by N. ¼ N.—From this to 10h 30m, bright ftreamers at intervals, low in the NW. quarter.— After 10h 30m grown faint; horizontal extent from WNW. to NE. by N.

April 14. At Kefwick, about midnight, or foon after, a ftill, horizontal light, altitude of the under edge 5°, of the upper edge 9°½; bearing of the centre NW. ¼ N.

L

the true folar time, it is prefumed this obfervation would be almoft cotemporary with that at *Kendal*.———Now, fuppofing this to be the cafe, the height of the *aurora*, or of the lower extremity of the beams, will be found equal to 62 Englifh miles.

GENERAL OBSERVATIONS.

IN order to determine the bearings of the middle or higheſt part of the arches of the *aurora*, I placed myſelf in a ſtation where I had a diſtant objeἀ before me, in the direἀion of the magnetic meridian, and I always found the higheſt part in the ſame direἀion as this objeἀ ;—a deviation of 2 or 3° would in moſt caſes have been very ſenſible. —Sometimes, to confirm the obſervation, equal altitudes of the arches were taken on each ſide of the magnetic meridian, with the theodolite, and the horizontal angle divided into two equal parts, which gave the ſame bearing of the centre as the other method.—It does not, however, always happen that the horizontal arch, eſpecially when high, is perfeἀ and complete.

The ſtreamers, or flaſhes, which pointed up, or perpendicular to the horizon, were only thoſe in the magnetic meridian, as well ſouth as north of the zenith.

The altitude of the centre of the *corona*, when there was one formed, was taken with a quadrant and plummet, with as much exaἀneſs as the thing ſeemed to admit of.

With regard to the needle of the theodolite, which was uſed to make the obſervations with, it is $3\frac{1}{2}$ inches long, and ſeems to move very freely upon its centre ; I have often tried the effeἀ of friἀion, by drawing it from its ſtation, and then ſuffering it to vibrate till it ſettled, when it uſually ſettled in the ſame ſtation within one or two minutes, but I have ſometimes obſerved it *five* minutes of a degree altered in ſuch a caſe.

I have never obſerved any conſiderable fluἀuation of the needle in any evening but when there was an *aurora* viſible, except once ; this was on the 13th of February, 1793, the evening of which was very wet and ſtormy ; the needle va-

ried

ried as follows :—the variation was 24° 57′ at noon; 24° 35′ at 5¼ P. M.; 24° 35′ at 5h 50m; 24° 20′ at 5h 58; 24° 20′ at 6h; 24° 48′ at 6h 20m; 24° 45′ at 6h 45m; 24° 35′ at 8h; 24° 47′ at 8h 30m; 24° 49′ at 10h 30m; 24° 53′ at 8 A. M. next day.

N. B. There had been an *aurora* the preceding evening.

It fhould alfo be noticed, that whilft making thefe ob-fervations upon the difturbance of the needle, during an *aurora*, I did not always know the *abfolute* variation at the time ; and therefore no inferences fhould be made relative to the change in the abfolute variation, in the interval from one *aurora* to another, from the obfervations I have given.

END OF THE FIRST PART.

METEOROLOGICAL

OBSERVATIONS AND ESSAYS.

PART SECOND.

ESSAYS.

ESSAY FIRST.

On the Atmosphere; its Constitution, Figure, Height, &c.

THE atmosphere is that invisible, elastic fluid which every where surrounds the earth, to a great height above its surface.—It was formerly supposed, that common air, or any portion of the atmosphere, when cleared of vapours and exhalations, was a pure, simple, elementary fluid; but modern philosophy has demonstrated the contrary, and it now appears that the purest air we breathe at any time, consists of an intimate

mixture

mixture of various elaftic fluids, or *gaffes*, in different proportions. Thofe properties of the atmofphere, called its falubrity and infalubrity, depend principally upon the greater or lefs quantity of one of its conftituent principles, *vital* or *dephlogifticated* air.—Whether the fuperior regions of the atmofphere confift in like manner of various elaftic fluids, or whether the fluids are the fame or different from thefe below, cannot, from the nature of the cafe, be determined experimentally.

The figure of the exterior furface of the atmofphere would, from the principles of gravitation, be fimilar to that of the earth, or of an oblate fpheroid ; or, its height and quantity of matter about the equator, would be fomething greater than at the poles, to preferve an equilibrium every where, owing to the centrifugal force, which is greateft at the equator. The denfity of the atmofphere, fuppofing it of an uniform temperature, and alike conftituted every where, would decreafe in afcending, in a geometrical progreffion : thus, if the denfity at one mile high was 1, and that at four miles high $\frac{1}{2}$; then that at feven miles high would be $\frac{1}{4}$, at ten miles high $\frac{1}{8}$, &c.——I fay thefe circumftances *would be*, were it not for the fun, or the principle of heat which it feems to produce ; but by means of the unequal diffufion of this principle, the circumftances are very materially different.

The

The mean annual temperature of the air, at the earth's furface, decreafes in going from the equator to the poles. Mr. *Kirwan* * ftates the mean annual heat at the equator at 84°, and that at the pole at 31°. Moreover, the temperature of the air over any place, in clear, ferene weather, decreafes in afcending above the earth's furface, nearly in an arithmetical progreffion, and at the rate of 1° for every hundred yards. Experience proves this, as far as to the fummits of the higheft mountains, which is about 3 miles; and hence it may be inferred to be fo above that height.

The great heat in the torrid zone rarefies the air, by increafing its elafticity; confequently the equilibrium of the atmofphere is difturbed. The rarefied air afcends into the higher regions, where, meeting with little refiftance, it muft flow northward and fouthward; the preffure upon the northern and fouthern regions is thus increafed, and a current muft fet in below, towards the equator, to reftore the equilibrium.— Hence, the higher temperature within the torrid zone, fwells the atmofphere there, and raifes it, or at leaft the grofs parts of it, to a much greater height than elfewhere; whilft in the frigid zone it is contracted by cold.—This is the effect of the different temperatures at the earth's furface: but the increafe of cold in afcending deftroys
the

* *Eftimate of the temperature of different Latitudes.*

the law of decreafe in denfity above mentioned, and greatly contracts the height of the atmof-phere, as deduced from fuch law; though this circumftance has perhaps no effect upon the fi-gure of the atmofphere.

Philofophers have attempted to find the height of the atmofphere by two methods; namely, by the duration of twilight, and by experiments upon the defcent of the barometer on high mountains. The former determines the height about 45 miles, as follows:—the twilight difap-pears when the fun is 18° below the horizon; hence it is argued, that a ray of light emitted from the fun, fo as to be a tangent to the earth's furface, after paffing through the atmofphere, is reflected from its external furface fo as to be a tangent to the earth's furface again, at 18° dif-tance from the former place of contact. This argument being admitted, affords *data* to find the height of the atmofphere, a proper allowance for refraction being firft made.—Several objec-tions to this conclufion however may be ftated; amongft others, it may be faid, we do not know whether the light, which comes to us at the dawn or departure of day, has been once or twice reflected; it may, and probably does, proceed from the zone of the earth illuminated by the twilight itfelf; in this cafe, therefore, we can determine no more from the twilight, than that the height of the atmofphere, or of that

region

region of it which is denfe enough to reflect light, is not fo much as 45 miles.

Barometrical experiments afford a much furer approximation to the height of the atmofphere or rather perhaps of the more grofs and heavy parts of it. From thefe we are affured, that a *ftratum* of air reaching from the earth's furface to the height of 4 Englifh miles, at all times contains above *one half* of the quantity of matter in the whole atmofphere; and by extending the laws thence refulting, we infer, that a *ftratum* 12 or 13 miles high, contains $\frac{3.8}{3.8}$ths of the whole: or, if a barometer, ftanding at 30 inches, was elevated to that height, the mercury would fall 29 inches.

The following table and theorem, extracted from Sir *George Shuckburgh*'s letter to Col. *Roy*, (Philofophical Tranfactions, Vol. 68.) will ferve to give my readers an idea in what manner the barometer is made fubfervient to the purpofe; and alfo how the height of mountains, &c. may be afcertained by means of the barometer.——In order to underftand the ufe of the table, it fhould be obferved, that two perfons are to take cotemporary obfervations, upon two barometers and thermometers, one perfon having one of each at the bottom of the moun-· tain, and the other at the top.

The

Thermo-meter.	Feet.
The Table.	
32°	86.85
35	87.49
40	88.54
45	89.60
50	90.66
55	91.72
60	92.77
65	93.82
70	94.88
75	95.93
80	96.99

EXPLANATION.

This table gives the number of feet in a column of the atmofphere, equivalent in weight to a like column of quickfilver $\frac{1}{10}$th of an inch high, when the barometer ftands at 30 inches, for every 5° of temperature from 32 to 80*.—For any other height of the barometer it will be in the inverfe ratio of that height to 30.——Let A = the mean height of the two barometers, in inches; a = the difference of the two, in tenths of an inch; b = the number of feet, per table, correfponding to the mean height of the two thermometers; x = the height of the mountain, in feet : then, we fhall have this theorem, $\frac{30ab}{A} = x$, the height required.

EXAMPLE.

Suppofe the barometer at the bottom to be 29.72 inches, thermometer 64°; the barometer at the top 27.46, thermometer 58°; required the height of the mountain?

Here the mean height of the two barometers, or A = 28.59 inches; their difference in tenths

M   of

---

* From the table it appears, that, in round numbers, every 30 yards of elevation reduces the height of the mercury in the barometer $\frac{1}{10}$ of an inch, near the earth's furface.

of an inch, or $a = 22.6$; the mean heat of the
two thermometers $= 61°$; the proportional
number may be found from the table $= 92.98$
feet $= b$; hence, $\dfrac{30 \times 22.6 \times 92.98}{28.59} = 2205$ feet,
the height required.

From this theorem we can deduce another:—
fuppofing the elevation of the upper barometer
given, and the height of its mercurial column
required; the other *data* as before.——Let H
$=$ the height of the barometer below, in inches;
$b =$ the number of feet, per table, as before*;
$p =$ the perpendicular elevation of the upper
barometer, in feet; $y =$ the height of its mercurial column, in inches: then, we obtain this
theorem, $y = \dfrac{600b - p}{600b + p} \times H$.

Hence we may calculate the height of the
mercurial column of the barometer at any given
moderate elevation, and by repeating the procefs, for a larger alfo, fufficiently accurate for
the purpofe of explaining the theory of the variation of the barometer; though we cannot
from this fix the boundary of the atmofphere
with precifion. To what height the very thin
and rare medium in the higher regions rifes,
we cannot afcertain; but there is fufficient reafon

* The height of the thermometer below being given, the
height of that fuppofed above may be eftimated, by deduct-
ing 1° for every hundred yards of elevation.

fon to conclude, as will be feen in a fubfequent Effay, that it extends to a much greater height than has commonly been fuppofed.

The following table contains the refult of a calculation from the laft mentioned theorem, of the height of the mercurial column, at certain elevations, above the equator, and likewife over the north of *England*, and the north pole. The mean heat at the earth's furface, under the e-quator, is fuppofed 84°; the mean heat in thefe parts, for the hotteft month of fummer, at 60°, and for the coldeft month of winter at 35°; the mean annual temperature at the north pole being fuppofed 31°, the mean temperature for the coldeft month of winter at that place may perhaps be ftated at 2°.

Elevation of the barometer above the level of the fea, in English miles.	Height of the mercurial column of the barometer, in inches.		
	Above the equator.	Above the North of England.		Above the north pole.
		In fummer.	In winter.	In winter.
0	30.00	30.00	30 00	30.00
2	20.55	20.10	19.58	18.81
4	13.61	12.96	12.24	11.19
6	8.66	7.98	7.26	6.24
8	5.25	4.65	4.03	3.19
10	3.00	2.52	2.05	1.45
12	1.58	1.24	.93	.56

ESSAY

ESSAY SECOND.

On Winds.

WINDS have ever been confidered, with reafon, as having a principal fhare in producing changes of weather, and therefore they demand a particular regard in meteorology.

Moft people know that the winds are not every where fo changeable as in thefe parts. In the torrid zone, the winds are much more uniform in direction than they are either in the temperate or frigid zones : over the Atlantic and Pacific oceans, particularly between 30° of north and 30° of fouth latitude, the *trade winds,* as they are called, blow pretty uniformly from eaft to weft, all the year round, with a fmall variation in the different feafons.

The caufe of thefe conftant winds, within the tropics, the ingenious and learned Dr. *Halley* has endeavoured to explain, and his explication feems to have been univerfally adopted by others fince its publication.—The chief phyfical prin- ciple he ufes, is the undeniable and well known one, that the air is rarefied by heat ; and, as the earth, in revolving from weft to eaft, expofes the torrid zone every day to the direct rays of

the

the fun, the earth, and confequently the air, is
there moft heated ; the *maximum* of heat follows
the fun, and therefore moves in a contrary direction, or from eaft to weft ; the rarefaction
occafioned thereby difturbs the equilibrium of
the atmofphere fucceffively ; and he argues, that
a current of air will conftantly follow the extreme of heat, to reftore the equilibrium,—and
thus he accounts for the trade winds.

It appears to me, however, that this conclufion
is premature, and not warranted by the laws of
motion. For, to fimplify the conception, let us
fuppofe a ring with a number of beads arranged
upon it at equal diftances, and, abftracting from
the force of gravity, that each of them is endued
with a repulfive power, in the fame manner as
are the particles of air. This fuppofition being
made, let the principle of heat, or any other
power, which acts fimply by increafing their
elafticity, act upon them in one part of the ring
more than in another; this will of courfe feparate the particles in fuch part farther than they
were before, and condenfe the others ; but it
can never produce a rotary motion of the whole
number of them round the ring, becaufe the action being mutual, the motion generated muft
be equal and contrary ;—or, in other words, no
momentum of the whole mafs of particles around
the ring, can be produced by any forces, which
they exert upon each other, agreeably to *New-*
ton's

ton's third law of motion.—We have here fuppofed the heat applied to one part of the ring
only, but it is plain the fame conclufion will obtain if it be applied to feveral parts at the fame
time, or fucceffively, or in any other manner;
likewife if the *addition* of heat produce no momentum, the *abftraction* of it will not.

Now to apply this to the matter in queftion :
let the fun be upon the equator, and the air un-
derneath be heated ; then the air in the plane of
the equator cannot recede from that plane, becaufe the lateral preffure on each fide will be
equal; and the action of the particles in the
faid plane upon each other, will be in the fame
circumftance as that of the particles upon the
ring, with refpect to any horizontal motion that
may be produced in the plane by the heat of the
fun.   It appears then, that no rotary motion of
the air round the earth can be produced by the
action of the fun upon the particles in that plane;
and by a like method of reafoning it may be
proved, that no fuch motion can be produced in
any other parallel plane ; confequently, the caufe
we are fpeaking of, or the fucceffive rarefaction
of the air from eaft to weft, cannot produce the
effect in queftion, nor immediately contribute
thereto.

It will be afked, if the trade winds are not
produced by the fucceffive rarefaction of the

parts

On Winds. 87

parts of the atmofphere within the torrid zone, what are they produced by?—To this it may be replied, that they admit of an explanation upon mechanical principles without requiring any hypothetical reafoning, or any other phyfical principle than that Dr. *Halley* ufes; namely, that heat rarefies the air. The inequality of heat in the different climates and places, and the earth's rotation on its axis, appear to me the grand and chief caufes of all winds, both regular and irregular; in comparifon with which all the reft are trifling and infignificant. The trade winds in the torrid zone, and the variable winds every where elfe, feem to be the natural effects of thefe two caufes, and might have been deduced from them *a priori*, if the facts had never been afcertained by the navigation of the torrid zone. Notwithftanding, as we are in poffeffion of many facts relative to the winds, it may be proper firft to ftate them, and then to confider how they refult from the caufes above mentioned.

Facts relating to the Winds.

1. Over the Atlantic and Pacific oceans, as has been obferved, the trade-winds extend from 30° of north to 30° of fouth latitude.

2. When the fun is on the equator, the trade-winds, in failing northward, veer more and more from the eaft towards the north; fo that about their limit they become nearly NE. :
and

and, *vice verfa*, in failing fouthward, they become at laft almoft SE.

3. When the fun is near the tropic of cancer,
the trade-winds north of the equator become
more nearly eaft than at other times, and thofe
fouth of the equator more nearly fouth : and,
vice verfa, when the fun is near the tropic of
capricorn.

4. The trade-wind is not due eaft upon the
equator, but about 4° to the north of it.

5. The winds in the northern temperate zone
are variable, but the moft general are the SW.
and W. and the NE. and E.——See page 48.

6. In the northern temperate and frigid zones,
and doubtlefs in the fouthern alfo, the winds are
more tempeftuous in winter than in fummer.——
See page 49.

Now in order to perceive the reafon of thefe
facts, it muft be remembered, that the heat is at
all times greateft in the torrid zone, and decreafes in proceeding northward, or fouthward;
alfo, that the poles may be confidered as the
centres of cold at all times : hence it follows,
that, abftracting from accidental circumftances,
there muft be a conftant afcent of air over the
torrid zone, as has been obferved, which afterwards falls northward and fouthward; whilft the
colder air below is determined by a continual
impulfe towards the equator. And, in general,
wherever the heat is greateft, there the air will
afcend,

afcend, and a fupply of colder air will be received from the neighbouring parts.———Thefe
then are the effects of the inequality of heat.

The effects of the earth's rotation are as follow : the air over any part of the earth's furface,
when apparently at reft or calm, will have the
fame rotary velocity as that part, or its velocity
will be as the co-fine of the latitude ; but if a
quantity of air in the northern hemifphere, receive an impulfe in the direction of the meridian,
either northward or fouthward, its rotary velocity will be greater in the former cafe, and lefs
in the latter, than that of the air into which it
moves ; confequently, if it move northward, it
will have a greater velocity eaftward than the air,
or furface of the earth over which it moves, and
will therefore become a SW. wind, or a wind
between the fouth and weft. And, *vice verfa,*
if it move fouthward, it becomes a NE. wind.
Likewife in the fouthern hemifphere, it will appear the winds upon fimilar fuppofitions will be
NW. and SE. refpectively *.

The trade-winds therefore may be explained
thus : the two general maffes of air proceeding
N from

---

* M. *De Luc* is the only perfon, as far as I know, who
has fuggefted the idea of the earth's rotation altering the
direction of the wind, which idea we have here purfued
more at large.—Vid. " Lettres phyfiques, &c." Tom. 5.
Part. 2. Let. cxlv.

from both hemifpheres towards the equator, as
they advance are conftantly deflected more and
more towards the eaft, on account of the earth's
rotation; that from the northern hemifphere,
originally a north wind, is made to veer more
and more towards the eaft, and that from the
fouthern hemifphere in like manner is made to
veer from the fouth towards the eaft; thefe two
maffes meeting about the equator, or in the torrid zone, their velocities north and fouth deftroy
each other, and they proceed afterwards with
their common velocity from eaft to weft round
the torrid zone, excepting the irregularities produced by the continents. Indeed the equator is
not the centre or place of concourfe, but the
northern parallel of 4°; becaufe the centre of
heat is about that place, the fun being longer on
the north fide of the equator than on the fouth
fide. Moreover, when the fun is near one of
the tropics, the centre of heat upon the earth's
furface is then nearer that tropic than ufual, and
therefore the winds about the tropic are more
nearly eaft at that time, and thofe about the
other tropic more nearly north and fouth.

Were the whole globe covered with water, or
the variations of the earth's furface in heat regular and conftant, fo that the heat was the fame
every where over the fame parallel of latitude,
the winds would be regular alfo: as it is, however, we find the irregularities of heat, arifing
from

from the interfperfion of fea and land, are fuch,
that though all the parts of the atmofphere in
fome fort confpire to produce regular winds
round the torrid zone, yet the effect of the fituation of land is fuch, that ftriking irregularities
are produced: witnefs, the monfoons, fea and
land breezes, &c. which can be accounted for
on no other principle than that of rarefaction;
becaufe the rotary velocity of different parallels
in the torrid zone is nearly alike.—For this reafon we have omitted giving the facts, and their
explanation, as having been done by others.

From what has been faid it might be fuppofed
that the winds in the northern temperate zone
fhould be between the north and eaft below,
and between the fouth and weft above, almoft
as regularly as the trade-winds; but when we
confider the change of feafons, the different capacities of land and water for heat, the interfe-
rence and oppofition of the two general currents,
the one of which is verging towards a central
point, and the other proceeding from it, we
might conclude it next to impoffible that the
winds in the temperate and frigid zones fhould
exhibit any thing like regularity: notwithftanding this, obfervations fufficiently evince, that the
winds in this our zone are, for the moft part,
in the direction of one of the general currents;
that is, fome where between the north and eaft,
or elfe between the fouth and weft; and that

winds

winds in other directions happen only as accidental varieties, chiefly in unfettled weather.

In winter, the heat decreafes more rapidly in
leaving the equator, and proceeding northward,
than at any other feafon ; confequently the currents of air to and from the equator, in the
northern hemifphere, move with the greateft velocity, and occafion the moft tempeftous weather, in that feafon : and, *vice verfa*, in fummer.

The effect of the earth's rotation to produce,
or rather to accelerate the relative velocity of
winds, being as the difference betwixt the co-
fines of any two latitudes, (or, to fpeak more
ftrictly, the effect is as the fluxion of the co-fine
of the latitude, the fluxion of the latitude being
fuppofed conftant) it will be fmall within the
torrid zone, and increafe in approaching the
poles. The hourly rotary velocity of the equator is about 1040 Englifh miles ; if we fuppofe
it 1000 miles it will be accurate enough for our
purpofe, and then, from a table of natural fines,
the rotary velocity of any parallel may be had at
once ; the differences of thefe velocities, will
ferve to give us fome idea of the comparative
effect of the earth's rotation at different parallels ;
for which purpofe we have fubjoined a table,
giving the rotary velocity of the parallels of latitude for every 10 degrees, together with their
differences, agreeable to the above fuppofition.

Degrees

Degrees of latitude.	Hourly rotary velocity of the parallels, in Englifh miles.	Differences of their velocities.
o		
o	1000	
10	984.8	15.2
20	939.7	45.1
30	866	73.7
40	766	100
50	642.8	123.2
60	500	142.8
70	342	158
80	173.6	168.4
90		173.6

From the table it appears, the effect of the earth's rotation, to accelerate the relative velocity of winds, is about ten times as great at the poles as at the equator;—by *relative velocity*, my readers will perceive I mean, all along, the velocity of the wind relative to the place of the earth's furface over which it blows ; hence, the relative velocity and direction of the mafs of air from the equator is at firft altered very flowly, and afterwards more rapidly, by the earth's rotation; and, *vice verfa*, with refpect to that from the poles.

Had the trade-winds been produced by the daily rarefaction of the air from eaft to weft alone, independent of the earth's rotation, they fhould have extended to 50° of north latitude when the fun is at the tropic of cancer, becaufe the heat at that parallel is then as great as at 30° of fouth latitude, which is quite contrary to experience : in fact, they ought to have extended,

tended, in a greater or lefs degree, over the ocean, from the equator to the poles, and the fummers have been more tempeftuous than the winters, becaufe the daily variation in heat is then greateft; neither of which we find confift-ent with obfervation.

The relative velocity of winds may be beft af-certained by finding the relative velocity of the clouds, which, in all probability, is nearly the fame as that of the winds; the velocity of a cloud is equal to that of its fhadow upon the ground, which, in high winds, is fometimes a mile in a minute, or 60 miles an hour; and a brifk gale will travel at the rate of 20 or 30 miles an hour. —It may be imagined, that the relative velocity of winds fhould be continually upon the increafe, by reafon that their caufes are conftantly in ac-tion, and not for a moment only; but the re-fiftance which a current of air meets with from the atmofphere itfelf, and from objects upon the earth's furface, muft be very confiderable; the increafe or diminution of the relative velocity of a wind will therefore depend upon the propor-tion between the active caufes and the refiftance.

The œconomy of winds, an illuftration of which we have been here attempting, is admi-rably adapted to the various purpofes of nature, and to the general intercourfe of mankind:— had the fun revolved round the earth, and not
the

the earth on its axis, the air over the torrid
zone, and particularly about the equator, would
have been in effect ftagnant; and in the other
zones the winds would have had little variation
either in ftrength or direction; navigation, in
this cafe, would have been greatly impeded, and
a communication between the two hemifpheres,
by fea, rendered impracticable. On the prefent
fyftem of things, however, the irregularity of
winds is of the happieft confequence, by being
fubfervient to navigation; and a general circulation of air conftantly takes place between the
eaftern and weftern hemifpheres, as well as between the polar and equatorial regions; by reafon of which, that diffufion and intermixture of
the different aerial fluids, fo neceffary for the
life, health, and profperity of the animal and vegetable kingdoms, is accomplifhed:—fuch is the
tranfcendent wifdom and providential care of the
common FATHER OF ALL!

PROOF OF THE EARTH'S ROTATION.

The trade-winds being matter of fact, if the
mechanical principles we have explained them
upon be admitted, we may draw from hence a
very fatisfactory, and indeed conclufive argument
for the earth's rotation on its axis; for, the
trade-winds blowing from eaft to weft, we muft
conclude, *a pofteriori*, that the earth revolves the
contrary way, or from weft to eaft.

ESSAY

ESSAY THIRD.

On the variation of the Barometer.

THE caufes of the variation of the barometer have never yet been difcovered, fo as to admit of demonftration; though feveral eminent philofophers have given the public the refult of their reafoning and experience on the fubject. We propofe to confider the principal of their allegations; but in the firft place it will be proper to lay down the chief *facts* refpecting the variation, which are the refult of obfervation, and not of any hypothefis.

Facts relating to the Barometer.

1. The barometer has very little variation within the tropics.

I believe the barometrical range has not been obferved much to exceed half an inch, in the torrid zone.

2. Within the northern temperate zone, and doubtlefs the fouthern alfo, the range of the barometer increafes in going from the equator.

The mean annual range* at *Paris*, in latitude 48° 50′ N. for 20 years, was 1½ inch; the greateft range, or difference between the higheft and loweft obfervations, for the fame term, was 2 inches. (Vid. *Martyn's Abridgment of the Parifian*

---

* By *annual range*, I mean the difference between the higheft and loweft obfervations each year.

fian Memoirs). At *Kendal*, in latitude 54° 17′ N. the mean range for 5 years was 2.13 inches; the greateft range was 2 65 inches. A comparifon of the obfervations made at *London, Kendal,* and *Kefwick* likewife corroborates the fame. —In *Sweden,* and *Ruffia,* the range is ftill greater.

3. In the temperate zones the range and fluctuation of the barometer is always greater in winter than in fummer.

See the obfervations, particularly the tables, p. 16 and 17.

4. The rife and fall of the barometer are not local, or confined to a fmall diftrict of country, but extend over a confiderable part of the globe, a fpace of two or three thoufand miles in circuit at leaft.

See the general obfervation, page 16.

In the French Philofophical Tranfactions for 1709, there is a comparifon of obfervations upon the barometer made at *Paris* and *Genoa,* for 3 years; the diftance of the places is at leaft 350 miles; notwithftanding this, it was found to rife and fall almoft univerfally on the fame day at both places, only the variation was lefs at *Genoa* than at *Paris,* becaufe its latitude is lefs; no difference in time was perceived, whether the fluctuations were fudden or gradual, except in one inftance, when the rife was one day later at *Genoa* than at *Paris.*

The precife extent to which the fluctuations of the barometer reach, has not, that I know of, ever yet been afcertained in any one inftance, for want of cotemporary obfervations made at a great number of diftant places.

5. The barometrical range is greater in *North America* than in *Europe,* in the fame latitude.

From the American Philofophical Tranfactions we find the range is as great in *New England* as in this country, though it is 10° nearer the equator. Alfo, at *Williamfburg,*

O   in

in *Virginia*, latitude 37° 20′ N. the annual range is above 1 inch, which is the fame as at *Genoa*, latitude 44° 25′ N.

6. In the temperate zones the mean ſtate of the barometer in the ſummer months is nearly equidiſtant from the extremes in that ſeaſon; but in winter the mean is much nearer the higher extreme than the lower.

According to the obſervations at *Kendal* (ſee page 16*) the mean height of the barometer in July is diſtant from the higher extreme 33 of an inch, and from the lower extreme .37; in January the mean is diſtant from the higher extreme .79, and from the lower 1.17: the ratio of the former diſtances is as 11 to 12, and of the latter as 8 to 12, nearly.

Profeſſor *Muſſchenbroek*, in his Elements of Natural Philoſophy, (tranſlated by *Colſon*) publiſhed about 50 years ago, has endeavoured to account for thoſe changes of weight in the atmoſphere; he has adverted to all or moſt of the cauſes that have ever been conſidered as agents in producing the effects: he enumerates the following cauſes, namely;—Firſt, the oppoſition of winds; ſecond, the north wind blowing, which cools and condenſes the air; third, the winds blowing upward or downward; fourth, an increaſe or diminution of heat, which rarefies or condenſes the air, in conſequence of which the air's diſtance from the earth's centre is increaſed or diminiſhed, and its weight, as well as centrifugal force, thereby affected; fifth, the air
being

* The mean for July, uncorrected, is 29.77, and for January 29.66, which muſt be uſed in this caſe, becauſe the extremes are not corrected.

being loaded with, or cleared of vapours and exhalations.

Profeſſor *De Sauſſure*, of *Geneva*, thinks the cauſes of the changes of the barometer are heat, different winds, and unequal denſity of the contiguous *ſtrata* of air; hence the little variation within the tropics. The principal cauſe is oppoſing winds. He does not deny that chymical changes in the air may affect the barometer; he however ſuſpects that ſome unknown cauſe has the greateſt effect *.———We ſhall now conſider the cauſes above alleged ſeverally.

The idea of oppoſite winds having the principal ſhare in producing the changes in the barometer, has evidently been ſuggeſted by the uniformity of the trade-winds, and the ſmall variation of the barometer where they blow; but it ſhould be conſidered, that the land-winds within the tropics do not always blow with the general or trade-winds, and that ſometimes they are in direct oppoſition; alſo, the monſoons, eſpecially about their change, produce uncommon conflicts of winds, and tempeſtuous weather, notwithſtanding which circumſtances, the barometer never has thoſe fluctuations that are experienced in the other zones. If, therefore,

the

* Theſe his ſentiments are taken from the *Critical Review,* for 1787.—Without being poſſeſſed of his work, we cannot examine his arguments particularly.

the idea of oppofite winds, mechanically accu-
mulating or difperfing the air, be inconfiftent
with the firft fact, it will certainly fail of explaining the reft. Befides, it would not be difficult to prove, *a priori*, that the oppofition of
winds, admitting the fact at the time, could not
produce thofe great and long continued accumu-
lations of air which we often experience.

The fecond caufe, or that of a cold north
wind blowing, has doubtlefs an effect upon the
barometer, though perhaps not altogether in the
manner that has been conceived.—We fhall confider this in another point of view by and by.

The third caufe, fuppofing it to exift at any
time, can only be local and tranfitory at moft;
but the rife or fall of the barometer is general,
and of confiderable duration : it cannot, therefore, produce the effect.

The fourth caufe is much too trifling to have
any material influence.

With refpect to the fifth, it muft be allowed,
that water, when changed into vapour, confti-
tutes a part of the atmofphere for the time, and
weighs with it accordingly ; alfo, that when vapour is precipitated in form of rain, the atmofphere lofes the weight of it : but it would be too
hafty to conclude from hence, that where evaporation is going forward the barometer muft rife,
and where rain is falling it muft fall alfo ; becaufe
air loaden with vapour is found to be fpecifically
lighter than without it.   Evaporation, therefore,
increafes

increafes the bulk and weight of the atmofphere at large, though it will not increafe the weight over any particular country, if it difplace an equal bulk of air fpecifically heavier than the vapour : and in like manner, rain at any place may not diminifh the weight of the air there, becaufe the place of the vapour may be occupied by a portion of air fpecifically heavier. It fhould feem therefore, that when the air over any country is cleared of vapours, &c. the barometer ought to be higher than ufual, and not lower. ——But we fhall now proceed to ftate our own ideas on the fubject.

It appears from the obfervations, (fee table, page 16) that the mean ftate of the barometer is rather lower than higher in winter than in fummer, though a ftratum of air on the earth's furface always weighs more in the former feafon than in the latter ; from which facts we muft unavoidably infer, that the height of the atmof- phere, or at leaft of the grofs parts of it, is lefs in winter than in fummer, conformable to the table, page 83. There are more reafons than one to conclude that the annual variation in the height of the atmofphere, over the temperate and frigid zones, is gradual, and depends in a great meafure upon the mean temperature at the earth's furface below ; for, clouds are never obferved to be above 4 or 5 miles high, on which account the clear air above can receive

little

little or no heat, but from the fubjacent regions
of the atmofphere, which we know are influenced
by the mean temperature at the earth's furface;
alfo, in this refpect, the change of temperature
in the upper parts of the atmofphere muft, in
fome degree, be conformable to that of the earth
below, which we find by experience increafes
and decreafes gradually each year, at any moderate depth, according to the temperature of the
feafon.   (See page 30.)

   Now with refpect to the fluctuations of the
barometer, which are fometimes very great in
24 hours, and often from one extreme to the
other in a week or 10 days, it muft be concluded,
either that the height of the atmofphere over
any country varies according to the barometer,
or otherwife that the height is little affected
therewith, and that the whole or greateft part
of the variation is occafioned by a change in the
denfity of the lower regions of the air.   It is
very improbable that the height of the atmofphere fhould be fubject to fuch fluctuations, or
that it fhould be regulated in any other manner
than by the weekly or monthly mean temperature of the lower regions; becaufe the mean
weight of the air is fo nearly the fame in all the
feafons of the year, which could not be if the
atmofphere was as high and denfe above the
fummits of the mountains in winter as it is in
fummer. However, the decifion of this queftion
need

need not reft upon probability; there are facts, which fufficiently prove, that the fluctuation of denfity in the lower regions has the chief effect upon the barometer, and that the higher regions are not fubject to proportionable mutations in denfity. In the memoirs of the Royal Academy at *Paris*, for 1709, there is a comparifon of ob- fervations upon the barometer at different places, and amongft others, at *Zurick*, in *Switzerland*, in latitude 47° N. and at *Marfeilles*, in *France*, latitude 43° 15′ N.; the former place is more than 400 yards above the level of the fea; it was found that the annual range of the baro- meter was the fame at each place, namely, about 10 lines; whilft at *Genoa*, in latitude 44° 25′ N. the annual range was 12 lines, or 1 inch; and at *Paris*, latitude 48° 50′ N. it was about 1 inch 4 lines. In the fame memoir it is related, that F. *Laval* made obfervations, for 10 days toge- ther, upon the top of *St. Pilon*, a mountain near *Marfeilles*, which was 960 yards high, and found that when the barometer varied 2¾ lines at *Mar- feilles*, it varied but 1¼ upon *St. Pilon*. Now had it been a law, that the whole atmofphere rifes and falls with the barometer, the fluctua- tions in any elevated barometer would be to thofe of another barometer below it, nearly as the abfolute heights of the mercurial columns in each, which in thefe inftances were far from being fo. Hence then it may be inferred, that the fluctuations of the barometer are occafioned

chiefly

chiefly by a variation in the denfity of the lower
regions of the air, and not by an alternate elevation and depreffion of the whole fuperincum-
bent atmofphere.   How we conceive this fluctuation in the denfity of the air to be effected, and
in what manner the preceding general facts relative to the variation of the barometer may be
accounted for, is what we fhall now attempt to
explain.

It has been obferved already that air charged
with vapour, or vapourized air, is fpecifically
lighter than when without the vapour; or, in
other words, the more vapour any given quantity of atmofpheric air has in it, the lefs is its
fpecific gravity.—M. *De Sauffure* has found from
experiment, that a cubic foot of dry air, of a
certain temperature, will imbibe 12 grains of
water; and that every grain of water diffolved
in air becomes an elaftic fluid capable of fup-
porting $\frac{1}{24}$ of an inch of mercury, while its denfity to that of air, is as 3 to 4 —Again, Dr.
Prieftley has found from frequent experiments
(vid. *Experiments and Obfervations relating to various branches of natural Philofophy*, Vol. 6, page
390) that different kinds of air, as for inftance,
inflammable air, and dephlogifticated air, the
fpecific gravities of which are as 1 to 12 nearly,
when mixed together, do not obferve the laws
of hydroftatics : for, the inflammable air, inftead
of rifing to the top of the veffel, diffufes itfelf
equally

equally and permanently through the dephlogifticated air, at the fame time that no chemical
attraction takes place betwixt them. The Doctor further obferves, " that the phlogifticated
" and dephlogifticated air, which compofe the
" atmofphere, are of very different natures,
" though without any known principle of at-
" traction between them, and alfo of different
" fpecific gravities; and yet they are never fe-
" parated but by the chemical attraction of fub-
" ftances, which unite with the one and leave
" the other."—Moreover, Sir *Benjamin Thomfon*
has found that moift air conducts heat better
than dry air. *(Vid.* Philofophical Tranfactions,
1786.)

From the two firft mentioned difcoveries we
may venture to infer, that if a cubic foot of dry
air were mixed with a cubic foot of moift air of
the fame temperature, the compound would oc-
cupy a fpace of two cubic feet, and be of equal
elafticity with the fimples, the two kinds of air
being intimately diffufed through each other.
Hence then a fluctuation of the denfity of the
air may happen thus : if a current of warm and
vapourized air flow into a body of cold and
dry air, it will difplace a part of the cold air, and
diffufe itfelf amongft the reft, by which means
the weight of the *ftratum* will be diminifhed,
whilft its bulk and fpring remain the fame ; and
vice verfa, if dry air flow into vapourized air.

P   The

The firſt faƈt mav then be accounted for thus:
—the warmer any air is, the more water it will
imbibe, in ſimilar circumſtances; hence, the air
over the torrid zone, being the hotteſt, will
contain the moſt vapour; and the air about the
poles, being the coldeſt, will contain the leaſt*:
moreover, as the heat within the torrid zone,
and the height of the atmoſphere there, remain
pretty nearly the fame all the year round, and
all the air approaching the zone from the two
temperate zones, is gradually aſſimilated in its
paſſage to that of the faid zone, it follows, that
there can be little fluƈtuation of denſity in the
lower regions of the air, and of courſe little variation of the barometer in the torrid zone.

The ſecond and third faƈts are the neceſſary
reſults of the principles we are aſſerting:—in
winter, the feafon when the barometrical range
is obſerved to be greateſt, the temperature of
the air decreaſes in proceeding from the torrid,
through the temperate, to the frigid zones; the
decreaſe

* The reader will pleaſe to obſerve, that the terms *moiſt
air*, and *vapourized air*, uſed in this and ſome other eſſays,
denote air containing a great portion of vapour, though it
mav not perhaps be charaƈterized as ſuch by a hygrometer.
—Thus, a cubic foot of air at the equator, which there is
indicated to be dry by a hygrometer, will contain more vapour than a cubic foot of air here, at the freezing temperature, which is indicated to be more moiſt than the former
by the hygrometer.—The difference of temperature produces this effeƈt.

decreafe is at firft moderate, but grows more
and more rapid as we advance; in confequence
of this decreafe, and the law by which it is
regulated, every place in the temperate zone
will, then more particularly, be fituate betwixt
the extremes of heat and cold, relative to its
own temperature, and the higher the latitude
the nearer will be thofe extremes to the place;
befides, that feafon being liable to the higheft
winds, the air will readily be transferred from
one parallel to another; and as the air at all
times will endeavour to maintain a proportion
of vapour fuitable to its temperature, it follows,
that the air in general in the higher latitudes
will then both be *cold* and *dry*, and in the lower
latitudes both *warm* and *moist*, relatively fpeaking. The confequence is obvious, that as a current from one or the other hand prevails, the
barometer will rife or fall accordingly and the
rife or fall will be greater as the place is fituate
nearer to the extremes of temperature, becaufe
the air will in that cafe fuffer the leaft change
in its paffage ——In fummer, the heat all over
the northern hemifphere is brought almoft to
an equality at the different parallels; the whole
mafs of air is heated, fwelled, and replenifhed
with vapour; the air over the northern regions is almoft brought into the fame ftate as
within the tropics, and the barometer therefore has almoft as little variation, in that feafon,
here as there.

P 2

The

The fourth fact offers nothing inconfiftent with our theory : winds are the mediate caufe of the variations of the barometer, and the currents of air to and from the torrid zone are not partial, but general, though fubject to confiderable modifications in direction ; befides, independent of winds, thofe properties of the air, heat and moifture, will always be diffufing themfelves in every direction, where there is a deficiency of either ; from which circumftances, it feems impoffible that the variations of the barometer fhould be local, though the amount of each fluctuation will not be the fame at places confiderably diftant. From the ufual celerity of the winds, the changes will happen upon the fame day at places very diftant ; but theory feems to require, that the northern parallels fhould firft experience the higher extremes, and the fouthern parallels the lower, and the obfervations upon the fourth fact countenance the inference. However, a feries of cotemporary obfervations made at two places, differing confiderably in latitude, would afcertain the fact ; and if the places were one NE. of the other, they would be ftill more eligible for the purpofe, becaufe the two general currents of air flow in that direction.

The climate of the eaftern coaft of *North America* is fo conftituted, that the decreafe of the mean temperature in the winter feafon, in proceeding northward, is much more rapid than

on

on the weftern coaft of this continent; the confequence is, that any particular place there is
liable to great and fudden fluctuations of temperature in that feafon, and thefe produce proportionate fluctuations of the barometer, according
as the warm and vapoury, or the cold and dry
air predominate.

The fixth fact has not, that I know of, ever
been accounted for, or even been adverted to,
by thofe who have attempted to explain the
caufes of the variation of the barometer; and
yet it will admit of a fatisfactory explanation
upon the principles we have adopted. Indeed,
at firft view, it feems inconfiftent with thofe
principles, becaufe we can produce no facts to
prove why the air may not deviate from its
mean ftate of heat and moifture as much towards one extreme as towards the other; but,
allowing what is moft probably the true ftate of
the cafe, that the deviations on each fide are
nearly equal, ftill the fact of the barometer admits of a rational folution.—Moift air, as has
been obferved, conducts heat much better than
dry air; now when the loweft extreme of the
barometer happens, the air is moift, high winds
generally prevail, and the atmofphere is much
ruffled by clouds and ftorms; all thefe circumftances tend to diffufe and circulate the heat, by
reafon of which the law of decreafe of temperature in afcending, at fuch times, muft be very
materially

materially different from what it is in ferene
weather; or, in other words, the decreafe of
temperature in afcending muft be much flower
than at other times; we may venture to fuppofe,
that, in fome cafes, the mean ftate of decreafe
for a few miles of elevation will be 1° for every
150 yards of afcent, inftead of 1° for every 100
yards, which is the ufual rate; the confequence
of this muft be a greater reduction of the barometer than otherwife would happen. For, let
the weight of the atmofphere at 3 miles of elevation be fuppofed equal to 15 inches of mercury,
the heat at the earth's furface equal to 45°, and
that it decreafes in afcending after the ufual rate
of 1° for every 100 yards; then, the mean heat
of a column of air from the earth's furface to 3
miles above it, will be 18°.6, whence the weight
of the whole column from the earth's furface to
the top of the atmofphere may be found by the
theorem, page 82; or $H = \frac{600b + p}{600b - p} \times y$ (y being given in this cafe) = 28.74 inches, the height
of the mercurial column of the barometer at the
earth's furface: but if we fuppofe the heat decreafes in afcending after the rate of 1° for 150
yards, then the mean heat of the column becomes equal to 27°.4, and the height of the
barometer equal to 28.30 inches; the difference
is .44 of an inch, occafioned by this change in
the temperature, which is greater by .06 of an
inch than the difference of the ranges above

and

and below the mean for January, at *Kendal,* as
ſtated at page 98.

The ſuppoſition made above, I preſume will
not be deemed extravagant, namely, that the
mean heat of a column of air 3 miles high will
not differ more from that at the earth's ſurface
than 17°, on certain occaſions: when we conſider the ſtrong SW. winds during a thaw,
(when the loweſt extreme uſually happens) and
that the thermometer often riſes to 45° at the
ſame time that the froſt is in the earth, and the
ground not cleared of ſnow, we muſt conclude,
that the then increaſing heat comes from the air
above, and not from the earth, and conſequently
that the temperature of the air is greateſt at a
conſiderable elevation, and decreaſes from thence
downward as well as upward; which circumſtance alone will greatly add to what the mean
temperature of the column would otherwiſe be.
—This irregularity and inverſion of the law of
heat in the atmoſphere, by which the loweſt extreme of the barometer is removed farther from
the mean ſtate than the higheſt, can only happen
in winter, by means of a ſudden influx of warm
air into cold; but in ſummer the heat of the air,
being chiefly derived from the earth's ſurface,
will be more equably diffuſed upwards, and
prevent ſuch a diſproportion in the diſtances of
the extremes from the mean, agreeably to obſervation.

Having

Having now endeavoured to explain the principal facts relative to the variation of the barometer, we shall next advert to some other particulars on the subject, which tend to illustrate and confirm the doctrine we have advanced.

The barometer generally rises with a wind betwixt the north and the east; it rises very high during a long and uninterrupted frost; it was highest for the last 5 years in January 1789; the mean temperature at *Kendal*, for 4 weeks preceding, was 28°, which was lower than for any other similar interval in the 5 years; there was only 1.643 inches of rain and snow for 7 weeks before; these were clear proofs of the prevalence both of cold and dry air.

The barometer is often low in winter, when a strong and warm S. or SW. wind blows; the annual extremes for these 5 years have always been in January; the lowest was in January, 1789, about 2 weeks after the above mentioned high extreme; it was accompanied with a strong S. or SW. wind, and heavy rain; the temperature of the air at the time was not high, being about 37°, but the reason was no doubt because one half of the ground was covered with snow; it was therefore probably warmer above.—Now the reason why the low extreme should have at that time, as well as at many others, soon succeeded the high extreme, seems explicable as follows :

follows: the extreme and long continued cold preceding, muſt have reduced the groſs part of the atmoſphere unuſually low, and condenſed an extraordinary quantity of dry air into the lower regions; this air was ſucceeded by a warm and vapoury current coming from the torrid zone, before the higher regions, the mutations of which in temperature and denſity are ſlow, had time to acquire the heat, quantity of matter, and elevation conſequent to ſuch a change below; theſe two circumſtances meeting, namely, a low atmoſphere, and the greateſt part of it conſtituted of light, vapoury air, occaſioned the preſſure upon the earth's ſurface to be ſo much reduced. Hence then, it ſhould ſeem, we ought never to expect an extraordinary fall of the barometer, unleſs when an extraordinary riſe has preceded, or at leaſt a long and ſevere froſt; this, I think, is a fair induction from the foregoing principles; how far it is corroborated by paſt obſervations, beſides thoſe juſt mentioned, I have not been able to learn.

It is obſervable that the high extreme ſome years happens in October or March, but generally in one of the intermediate months; the low extreme is moſtly in December or January. From the obſervations at *Paris* for 20 years, from 1699 to 1718, incluſive, if we take 11 of the loweſt that were made, 10 of them were in December and January, and the eleventh in November.

Q   The

The month of January, 1791, will be long remembered, on account of the loſſes at ſea, and damage at land, by the extraordinary high winds, which prevailed almoſt inceſſantly throughout the month, from the SW.—See page 49   Now a ſtrong and warm SW. wind blowing continually in that ſeaſon, when the atmoſphere was low, ought to have reduced the mean ſtate of the barometer unuſually low ; the faɛt therefore may be produced, as an *experimentum crucis* of the theory ; accordingly, we find from the obſervations, that the mean ſtate of the barometer for that month was lower by .14 of an inch, in the north of *England*, and probably lower every where on the weſtern coaſt of *Europe*, than for any other month in the laſt 5 years.

It does not appear from the barometrical obſervations in the firſt part of this book, that cold alone, independent of every other circumſtance, has a tendency to increaſe the mean weight of the atmoſphere over any place ; for, if it had, the mean ſtate of the barometer would be higher in winter than in ſummer, contrary to experience ; if, therefore, the mean ſtate of the barometer be lower in the torrid than frigid zones, it is moſt probably effeɛted by the vapoury air.

ESSAY

ESSAY FOURTH.

On the relation between Heat and other Bodies.

WE have nothing new to offer on this fubject; but as fome knowledge of the
matter is requifite in order to underftand fome
of the phenomena of meteorology, we purpofe
to give a brief explanation of fuch facts as may
be adverted to in the courfe of this work.

Different bodies that are equal in *magnitude*,
and of the fame temperature, do not contain
equal quantities of fire; neither do different bodies, that are equal in *weight* and temperature,
contain equal quantities of fire.—For example,
if a cubic inch of *iron* be heated to 100°, and
then thrown into a given quantity of water at
50°, the temperature of the water will be augmented; but if inftead of *iron, lead* be ufed, the
temperature will not be fo much augmented;
on the contrary, if the iron and lead were colder
than the water, the iron would diminifh its temperature moft. If equal *weights* of iron and lead
were ufed, the refults would be fomewhat different, but ftill the temperature of the water would
be more augmented or diminifhed by the iron
than by the lead. When equal *weights* are ufed
in experiments of this fort, that body which augments or diminifhes the temperature the moft, is
faid to have the greater capacity for heat; be-

Q 2   caufe

caufe a greater quantity of heat is required to be added to, or fubtracted from it, in order to vary its temperature equally with the other.

The fame body, under the different forms of *folid, fluid,* and *aeriform,* has different capacities for heat; in the folid form its capacity is leaft, and greateft in its aeriform ftate; alfo, when any folid body is converted into a non-elaftic fluid, or any non-elaftic fluid into an elaftic fluid, by heat, it abforbs a portion of heat during its converfion, which does not increafe its temperature; and when the change takes place the contrary way, by cold, it parts with an equal portion of heat, without having its temperature diminifhed. —To inftance in *ice, water,* and *aqueous vapour:* if a pound of ice were taken of the temperature of 20°, and a quantity of heat added to it, fo as to augment its temperature to 25°; an equal quantity of heat would augment the temperature of a pound of water lefs than 5°, and of aqueous vapour ftill lefs. Again, if a pound of ice of 32°, and a pound of water of 172° were mixed together, the temperature of the mixture would be 32°, becaufe the ice requires 140° of heat to melt it; that is, it requires as much heat to melt it as would increafe the temperature of a pound of water 140°; whereas, if a pound of water of 32° were mixed with a pound of water of 172°, the temperature of the mixture would be the mean betwixt the two, or 102°. Alfo, it has been found, that aqueous vapour, when con-

denfed

denfed into water of the fame temperature, gives
out 943° of heat.

The capacities of *earth, ftones,* and *fand,* for
heat, are much lefs than that of water. This is
one caufe why the viciffitudes of temperature
are greater at land than at fea*.

Another particular relative to heat is, that
fome bodies conduct it better than others; in
this refpect there is a ftriking refemblance between the electric fluid and fire; for, thofe bodies which conduct the electric fluid well, as
metals, water, &c. alfo conducts heat well.—
Glafs, fealing-wax, and other electrics, conduct
heat very flowly; alfo *dry land,* whether the furface be ftony, fandy, or earthy, is found by experience to conduct heat flowly.

Sir *B. Thomfon* has by a feries of experiments
(fee Philofophical Tranfactions, 1786) found the
powers of a few bodies to conduct heat to be
proportionate to the following numbers, namely:

Mercury - - - - - - - -	1000
Moift air - - - - - - - -	330
Water - - - - - - - -	313
Common air, denfity 1 - - - -	80.11
Rarefied air, denfity $\frac{1}{4}$ - - - -	80.23
Rarefied air, denfity $\frac{1}{24}$ - - - -	78
Torricellian vacuum - - - - -	55

E8SAY

* Thofe who wifh to fee the fubject touched upon above,
difcuffed at large, may perufe Dr. *Crawford's Experiments
and Obfervations on Animal Heat and the Inflammation of combuftible Bodies.*

ESSAY FIFTH.

On the Temperature of different Climates and Seasons.

MR. *Kirwan* has treated of this subject in so able a manner, that we can do little more than extract from his work *.

That the sun is the primary cause of heat all over the earth, is almost too apparent ever to have admitted of doubt; though some philosophers have imagined a *central heat* or body of fire in the earth, which, by its emanations, mitigates the severity of the winters in the higher latitudes: the opinion is, however, disproved by facts, which shew, that the temperature of places 30, 40, or 50 feet below the earth's surface, remains nearly the same all the year round as the mean annual temperature at the surface, and that at a less depth the temperature varies, in a small degree, with the season. The fact seems to be, that in winter the earth gives out to the atmosphere a portion of heat received in summer.

The earth's surface is the chief medium by which the sun heats the atmosphere; for it is observable

* *Estimate of the Temperature of different Latitudes.*

obfervable that clear air is not heated in any fenfible degree by the action of the fun's rays. The direct rays of the fun falling upon *ftony* or *fandy* ground, are found to increafe its temperature amazingly, partly on account of its fmall capacity for heat; whilft the temperature of water is thereby increafed very little, from its great capacity for heat, the reflection from its furface, and evaporation. Water being a much better conductor of heat than land, preferves a greater uniformity of temperature; whilft land is more fubject to the viciffitudes of heat and cold.

Living vegetables alter their temperature very flowly; the evaporation from their furfaces is much greater than from the fame fpace of land uncovered with vegetables: *forefts* prevent the fun's rays from reaching; hence, wooded countries are colder than thofe open and cultivated.

Evaporation and the condenfation of vapour are made fubfervient to the more equal diffufion of heat over the different climates and places: evaporation being great in the torrid zone, a vaft portion of heat is thereby abforbed, and rendered infenfible, till being carried northward or fouthward, the vapour is condenfed, and gives out its heat again, which being diffufed in the atmofphere, augments its temperature very confiderably.

Mr.

Mr. *Kirwan*, confidering thefe and other cir-cumftances, judges it moft eligible, in comparing the temperature of different places, to fix upon a fituation that may ferve as a ftandard of compa-rifon, and he judicioufly prefers the fea to the land, as being more free from accidental vari-ations. By combining theory with obfervation, he obtains the mean annual heat of the equator equal to 84°, and that of the pole 31°; and then gives the following theorem for the mean annual temperature of the ftandard fituation in every latitude; namely, if $S =$ the natural fine of any latitude to radius 1; then, $84 - 53 \times S^2 =$ the mean annual temperature of that latitude.

This theorem gives the temperature of dif-ferent latitudes as by the following table.

Table of the mean annual temperature of the ftandard fituation, for every 5 degrees of latitude.

Lat.	temp.	Lat.	temp.	Lat.	temp.	Lat.	temp.	Lat.	temp.
°	°	°	°	°	°	°	°	°	°
0	84	20	77.8	40	62.1	60	44.3	80	32.6
5	83.6	25	74.5	45	57.5	65	40.5	85	31.4
10	82.3	30	70.7	50	52.9	70	37.2	90	31
15	80.4	35	66.6	55	48.4	75	34.6		

It afterwards becomes neceffary to confider the modifications of the ftandard temperature on land, from fituation, &c.

1. Elevation diminifhes the mean temperature of places. Its effects Mr. *Kirwan* ftates as fol-lows:

lows: if the elevation be moderate, or at the rate of 6 feet per mile from the neareſt ſea; then, for every 200 feet of elevation, allow $\frac{1}{4}$ of a degree for the diminution of the mean annual temperature.

If the elevation be 7 feet per mile, allow $\frac{1}{3}$ of a degree.
    13 feet $\frac{4}{10}$
    15 feet, or upwards $\frac{1}{2}$

N. B. The elevation of any inland place may be found ſufficiently exaĉt for this purpoſe, by obſerving how much the mean annual height of the barometer falls ſhort of 30 inches, and allowing for the difference, according to the theorem in page 81 ; becauſe the mean annual height of the barometer, on a level with the ſea, is nearly 30 inches every where.

2. Next to elevation, diſtance from the ſtand-ard ocean ſeems to have the moſt conſiderable effeĉt upon the mean annual temperature ; its amount Mr. *Kirwan* ſtates, from a compariſon of obſervations, as follows: namely, the mean annual temperature is depreſſed or raiſed, for every 50 miles diſtance, nearly at the following rate :

From lat. 70° to lat. 35° *cooled* $\frac{1}{3}$ of a degree.
    35 to   30 ——— $\frac{1}{8}$
    30 to   25 *warmed* $\frac{1}{5}$
    25 to   20 ———$\frac{1}{2}$
    20 to   10 ———1°.

R   This

This effect of diſtance from the ſtandard ocean Mr. *Kirwan* ſeems to attribute to the unequal capacities of land and water for heat; but, with deference to the opinion of ſo reſpectable a philoſopher, I think this alone inadequate to the effect. For, if land in general receive *more* heat immediately from the ſun in a year than water, the mean temperature of the internal parts of the continent ought to be the greateſt from the equator to the pole. And if land receive *leſs* heat, then, for ought that appears, the mean temperature of the internal parts of the continent might be expected the leaſt in every latitude; but in neither caſe, I think, could we conclude *a priori*, from the mere difference of capacity, that the mean heat of the internal parts of the continent would be greater near the equator, and leſs more northward, than the mean heat upon the coaſt.—To account for the effect in queſtion, we ſhall therefore propoſe the following theory.

Let it be firſt ſuppoſed that water receives a greater quantity of heat, from the ſun's rays, than land in general, under every parallel of latitude*; in the next place, it will be allowed, that a much greater quantity of water is evaporated

* It is generally allowed, I think, that land reflects more light than water, and conſequently imbibes leſs; and the quantity of heat received will doubtleſs be proportionate to the rays imbibed.

rated from the fea, in the torrid zone, than from an equal area of land in the fame zone ; hence it will follow, that the quantity of heat abforbed by the vapour may, for ought we know, be fo great as to reduce the mean temperature of the fea there below that of the land : in fuch cafe it is evident, the further any place is diftant from the fea, the greater muft its mean temperature be, all other circumftances being the fame. Again, the farther we proceed northward, the lefs is the quantity of water annually evaporated from a given furface of the fea ; hence there may be a parallel of latitude where the heat abforbed by the greater evaporation of the fea, is equal to the heat which the fea receives more than the land ; in this cafe therefore, the mean temperature of the land and fea will be every where the fame in the fame parallel. Farther than this, the mean temperature of the fea will become greater than that of the land, and the more fo as the latitude increafes. It appears then, that the difference of the capacity of land and water for heat, requires to be joined to the fuppofition that water receives more abfolute heat than land from the fun's rays, before we can produce, *a priori*, a refult fimilar to what is ftated above as deduced from obfervation.

But if we purfue the thought ftill farther, we fhall perhaps find, that the above ftatement of the effect of diftance from the ftandard ocean,

is

is not altogether compatible either with theory
or obfervation,—and at the fame time draw a
conclufion of much importance to the fubject we
are now difcuffing.

It is obfervable, that in the northern temperate zone, the internal parts of the eaftern continent are generally hotter, in fummer, than on
the coaft under the fame parallel, except elevation or fome peculiarity of foil or fituation diminifh the temperature; but the cold of winter is
fo much more fevere, that the mean temperature
is greatly reduced below the ftandard — Now in
winter, when the influence of the fun is fo weak,
it fhould feem that the condenfation of vapour
alone affords the northern atmofphere a very
large portion of the fenfible heat it has in that
feafon. And it appears in the former effay on
winds, that the general current of air from the
equator is SW. when it arrives in the northern
temperate zone; this current coming from the
fea to the weftern coaft of each continent, will
there meet with cold air, which condenfes its
vapour as it proceeds, affording plenteous rain
and heat to the weftern coafts: as the current
proceeds into the internal parts of the continents
it lofes its vapour and heat, till at length the
precipitation becomes much lefs in quantity, and
in form of fnow; the current then continues its
progrefs, and grows colder and colder till it arrives at the eaftern coaft, unlefs the influx of fea
breezes

breezes mitigate the temperature near the coaſt. Hence then it may be inferred, *that in the temperate zones, the weſtern coaſts of all continents and large iſlands, will have a higher mean temperature than the eaſtern coaſts under the ſame parallel, and particularly will have more moderate winters.*

It remains now to ſhew how far this inference is countenanced by obſervatión.—We are certain that the eaſtern coaſt of *Aſia* is much colder than the weſtern coaſt of *Europe;* on the eaſtern coaſt of *Kamſchatka,* in latitude 55° N. Capt. *Cook* found ſnow 6 or 8 feet deep, in May, and the thermometer was moſtly 32°; and in January the cold is ſometimes — 28°, and generally — 8°. At *Pekin,* in *China,* latitude 39° 54´ N. longitude 116° 29´ E. the mean temperature is only 55°.5, the *Atlantic* under this parallel being 62°; the uſual range of the thermometer each year is from 5° to 98°, not unlike what it is at *Philadelphia,* which is under the ſame parallel. —Again, we are certain that the eaſtern coaſt of *North America* is 10 or 12° colder than the oppoſite weſtern coaſt of *Europe;* and hence it may be preſumed, that the weſtern coaſt of *North America,* or that of *California,* is warmer than the eaſtern. The NE. parts of *Siberia* on the one continent, and the country about *Hudſon's Bay* on the NE. ſide of the other continent, ſeem equally ſubjeċt to the moſt rigorous cold

in

in winter.——But to proceed to the other modifications of the ſtandard temperature.

3. As for the effects of mountains, foreſts, ſeas, &c. upon the mean annual temperature of places, Mr. *Kirwan* obſerves, that all countries lying to the windward of high mountains, and extenſive foreſts, are warmer than thoſe lying to the leeward, in the ſame latitude. Countries that lie ſouthward of any ſea are warmer than thoſe that have that ſea to the ſouth of them. Iſlands participate moſt of the temperature of the ſea, and are therefore not ſubject to the extremes of heat and cold ſo much as continents.

———

We ſhall here introduce a table containing the mean annual temperature of ſeveral places, as determined by obſervation, from the " Eſtimate, &c." page 113.

	North Lat.	Longitude.	Mean annual Feat.
Wadſo, in Lapland - -	70° 5		36°
Abo - - - -	60 27	22°18´ E.	40
Peterſburgh - - -	59 56	30 24 E.	38 8
Upſal - - - -	59 51	17 47 E.	41.88
Stockholm - - -	59 20	18 E.	42.39
Solyſkamſki - - -	59	54 E.	36.2
Edinburgh - - -	55 57	3 W	47.7
{ Keſwick* - - -	54 33	3 3 W.	46
{ Kendal - - - -	54 17	2 46 W.	46.4

Franeker

* Theſe two places are inſerted from page 29 and 30 of this work.

	North Lat.	Longitude.	Mean annual Heat.
Franeker - - -	53° 0	5°42' E.	52°.6
Berlin - - - -	52 32	13 31 E.	49
Lyndon, in Rutland -	52 30	0 3 W.	48.03
Leyden - - - -	52 10	4 32 E.	52.25
London - - -	51 31		51.9
Dunkirk - - - -	51 2	2 7 E.	54 9
Manheim - - -	49 27	9 2 E.	51.5
Rouen - - - -	49 26	1 W.	51
Ratisbon - - -	48 56	12 5 E.	49.35
Paris - - - -	48 50	2 25 E.	52
Troyes, in Champaigne -	48 18	4 10 E.	53.17
Vienna - - - -	48 12	16 22 E.	51.53
Dijon - - - -	47 19	4 57 E.	52.8
Nantes - - - -	47 13	1 28 E.	55.53
Poitieres - - -	46 39	0 30 E.	53.8
Laufanne - - -	46 31	6 50 E.	48.87
Padua - - - -	45 23	12 E.	52.2
Rhodes, in Guienne - -	45 21	2 39 E.	52 9
Bordeaux - - -	44 50	0 36 W.	57.6
Montpelier - - -	43 36	3 73 E.	60 87
Marfeilles - - -	43 19	5 27 E.	61.8
Mont Louis, in Roufillon -	42	2 40 E.	44.5
Cambridge, in New England	42 25	71 W.	50.3
Philadelphia - - -	39 56	75 9 W.	52.5
Pekin - - - -	39 54	116 29 E.	55.5
Algiers - - - -	36 49	2 17 E.	72
Grand Cairo - - -	30	31 23 E.	73
Canton - - - -	23	113 E.	75.14
Tivoli, in St. Domingo -	19		74
Spanifh Town, in Jamaica -	18 15	76 38 W.	81
Manilla - - - -	14 36	120 58 E.	78.4
Fort St. George - -	13	87 E.	81.3
Ponticherry - - -	12	67 E.	88
	South Lat.		
Falkland Iflands - -	51° 0'	66 W.	47 4
Quito - - - -	0 13	77 50 W.	62

The hotteft place mentioned in this table is *Ponticherry;* the heat there is fometimes 113 or 115°, which far exceeds that of the human body. The mean heat of June is 95°.4.

In

In fome parts of *Africa* the heat even exceeds that of *Pondicherry.*

Of all inhabited countries, *Siberia* feems the coldeft; its great elevation and diftance from the ocean both confpire to make it fo. Mercury has often been frozen there by the natural cold, which confequently exceeded — 39°. The mean temperature of *Irkutz*, latitude 52° 15′ N. longitude 105° E. from October 1780 to April 1781, was — 6°.8.

At *Peterfburgh* the cold has been known — 39°: and is one year with another, at an average, — 25°; the greateft fummer heat, on a mean, is 79°, yet once it amounted to 94°.

General Obfervations and Inferences.

Eftimate, &c. page 19. " The temperatures of different years differ very little near the equator, but they differ more and more, as the latitudes approach the poles.

" It fcarce ever freezes in latitudes under 35°, unlefs in very elevated fituations; and it fcarce ever hails in latitutes higher than 60°.

" Between latitudes 35° and 60°, in places adjacent to the fea, it generally thaws when the fun's altitude is 40°, and feldom begins to freeze until the fun's meridian altitude is below 40°."

Page 28. " The greateft cold, within the 24 hours, generally happens half an hour before fun-rife, in all latitudes. The greateft heat in
all

all latitudes between 60° and 45°, is found about half paſt 2 o'clock in the afternoon; between lat. 45° and 35°, at 2 o'clock; between lat. 35° and 25°, at half paſt 1; and between lat. 25° and the equator, at 1 o'clock.

" On ſea, the difference between the heat of day and night, is not ſo great as on land, particularly in low latitudes.

" The coldeſt weather, in all climates, generally prevails about the middle of January, and the warmeſt in July, though, aſtronomically ſpeaking, the greateſt cold ſhould be felt at the latter end of December, and the greateſt heat in the latter end of June; but the earth requires ſome time to take, or to loſe the influence of the ſun, in the ſame manner as the ſea, with reſpect to tides, does that of the moon."

Page 104, *&c.* " July is the warmeſt month in all latitudes above 48°; but in lower latitudes Auguſt is generally the warmeſt.

" December and January, and alſo June and July, differ but little. In latitudes above 30°, the months of Auguſt, September, October, and November, differ more from each other, than thoſe of February, March, April, and May. In latitudes under 30°, the difference is not ſo great. The temperature of April approaches

S   more,

more, every where, to the annual temperature, than that of any other month : whence we may infer, that the effects of natural caufes, that operate gradually over a large extent, do not arrive at their *maximum*, until the activity of the caufes begins to diminish ; this appears alfo in the operation of the moon on feas, which produces tides ; but after thefe effects have arrived at their *maximum*, the decrements are more rapid, than the increments originally were, during the progrefs to that *maximum* *.

" The differences between the hotteft and coldeft months, within 20° of the equator, are inconfiderable, except in fome peculiar fituations ; but they increafe in proportion, as we recede from the equator.

In

* The foregoing obfervations, made at *Kendal* and *Kefwick*, afford fome remarkable exceptions to the three laft general obfervations —December is the coldeft month in thefe places ; though perhaps a mean of 5 years is not fufficient to determine the point. Auguft is generally the warmeft month, and not July ; the reafon of this laft I take to be, our mountains being topped with fnow during the fpring, which retards the increafe of temperature, and throws the *maximum* of heat later in the fummer. For the fame reafon, the month of April is colder than the annual mean : October feems the neareft to it. The ftandard temperature for thofe places is 49° ; the difference, being between 2 and 3°, muft be attributed, I think, chiefly to the extenfive ranges of mountains and high lands, in almoft every direction ; unlefs, perhaps, we have determined the temperature too low.——See the obfervations, page 30.

In the higheft latitudes, we often meet with a heat of 75 or 80°; and particularly in latitudes 59° and 60°, the heat of July is frequently greater than in latitude 51°.

Every habitable latitude enjoys a heat of 60° at leaft, for 2 months; which heat feems neceffary, for the growth and maturity of corn. The quicknefs of vegetation, in the higher latitudes, proceeds from the duration of the fun over the horizon. Rain is little wanted, as the earth is fufficiently moiftened by the liquefaction of the fnow, that covers it during the winter; in all this, we cannot fufficiently admire the wife difpofition of Providence.

ESSAY

ESSAY SIXTH.

On Evaporation, Rain, Hail, Snow, and Dew.

EVAPORATION is that procefs in nature
by which water and other liquids are abforbed into the atmofphere, or are converted
into elaftic fluids, and diffufed through the atmofphere; the liquid thus changed, is termed
vapour, and the vapour is characterized by the
name of the liquid from which it was generated,
as *aqueous vapour*, or the vapour derived from
water, &c.—Whether the vapour of water is
ever chymically combined with all or any of the
elaftic fluids conftituting the atmofphere, or it
always exifts therein as a fluid *fui generis*, diffufed amongft the reft, has not, I believe, been
clearly afcertained.

The following circumftances are found pow-
erfully to promote evaporation; namely, *heat,
dry air*, and a *decreafed weight* or *preffure* of the
atmofphere upon the evaporating furface. The
firft and fecond are known to have that effect,
from every one's experience; the laft is proved
to have fuch an effect, by the air-pump. For,
when the air is exhaufted out of a receiver, a
large quantity of vapour is raifed from the wet
leather upon the pump plate; this vapour is
precipitated again when the air is let in, fo as to
appear

appear falling like a fhower*. If a quantity of
warm water be placed under a receiver, when
the air is rarefied to a fufficient degree, the water
boils with great violence, and a large portion of
it may in this manner be readily raifed in vapour,
which is as foon condenfed by the cold of the
furrounding medium, and falls upon the leather
of the pump-plate. The reafon of this is, that
the greateft heat water is fufceptible of, or its
boiling heat, depends upon the preffure of the
air upon its furface; the lefs the preffure, the
lefs is the boiling heat; and whenever it arrives
at the boiling heat, the greater heat applied to
augment its temperature, inftead of doing fo,
converts a portion of it into vapour, which, as
has been remarked, abforbs a great quantity of
heat, without any increafe of temperature†.

As this variation of temperature in boiling
water according to the different preffure of the
air, is a circumftance not foreign to the fubject
we are upon, and perhaps the quantity and mode
of the variation may not be generally known,
we fhall here introduce the refult of a feries of
experiments made in order to afcertain what
preffure upon the furface of water is requifite
to make it boil at a given temperature; having
never feen any fimilar account, though the thing
has

* See a note upon this fubject, page 136.

† Hence we fee the reafon of the provifo, page 20, in
determining the boiling point of thermometere.

has probably been done by others with more accuracy.

Heat of the water when boiling.	Preffure upon its furface, in inches of mercury.	Rarefaction of the air.
212°	30.00	1
200	22.8	1.3
190	18.6	1.6
180	15 2	2
170	12.2	2.45
160	9.45	3.2
150	7.48	4
140	5.85	5.1
130	4.42	6.8
120	3 27	9.2
110	2.52	11.9
100	1.97	15.2
90	1.47	20.4
80	1.03	29

N. B. M. *De Sauffure* found the heat of boiling water upon the fummit of mount *Blanc,* 186°; the height of the mountain is near 3 miles above the level of the fea; the barometer was 16 inches $\frac{144}{160}$ of a line (a little above 17 Englifh inches.

Experiments of this fort, when made with all the accuracy they will admit of, I am inclined to think will lead to the true theory of evaporation, and to the ftate of vapour in the atmofphere; upon confideration of the facts, it appears to me, that evaporation and the condenfation of vapour are not the effects of chymical affinities, but that aqueous vapour always exifts as a fluid *fui generis,* diffufed amongft the reft of the aerial fluids.

fluids.—It is true, the fact that a quantity of common air of a given temperature, confined with water of the fame temperature, will only imbibe a certain portion of the water, and that the portion increafes with the temperature, feems characteriftic of chymical affinity; but when the fact is properly examined, it will, I think, appear, that there is no neceffity of inferring from it fuch affinity.

Granting the truth of the preceding experiments, when the incumbent air is rarefied 29 times, water of 80° is at the point of ebullition; or, in other words, aqueous vapour of the temperature of 80°, can bear no more than 1.03 inches of mercury, without condenfation; this, then, is the extreme denfity of the vapour of that temperature. Now, when a quantity of atmofpheric air of 80° imbibes vapour, the vapour is diffufed through it, and it may therefore continue to imbibe till the denfity of the vapour, confidered abftractedly, becomes $\frac{1}{29}$ of what it is when under the preffure of 30 inches of mercury, and its temperature 212°; or, till $\frac{1}{29}$ of the bulk of the compound mafs is vapour, and then it will be faturated, or imbibe no more; becaufe if it did, the denfity of the vapour muft be increafed, which it cannot be in that temperature, without lofing its form, and becoming water. Thus then it appears, that upon this hypothefis, there is no need to fuppofe a chy-

mical

mical attraction in the cafe ; and further, that a
cubic foot of dry air, whatever its denfity be,
will imbibe the fame weight of vapour if the
temperature be the fame ; and laftly, that it may
be determined *a priori*, what weight of vapour
a given bulk of dry air will admit of, for any
temperature, provided the fpecific gravity of the
vapour be given.   For example, let it be required to find the weight of vapour which a
cubic foot of dry air of 80° will admit of, or imbibe, fuppofing the fpecific gravity of air .0012,
and that of vapour to air as 3 to 4 :—A cubic
foot of water weighs 437500 grains, and the
fpecific gravity of vapour from the *data*, is .0009;
now the compound mafs being denoted by q, we
fhall have $\frac{1}{29}q =$ the vapour, and $q = 1$ foot $+$
$\frac{1}{29}q$; that is, $q = \frac{29}{28}$ foot ; and the vapour $=$
$\frac{1}{28}$ foot, $= 14$ grains.   This, it will be obferved,
is the refult of the *hypothefis*.   M. *De Sauffure*
determined by the *experiment* alluded to, page
104, that a cubic foot of dry air of 66° would
imbibe 11 or 12 grains of water.   Hence then
it feems probable that the hypothefis would a-
gree with experiment.—By a like procefs, we
fhall find the weight of vapour imbibed by a cu-
hic foot of air of 150°, equal to 131 grains. *

Evaporation

* I cannot forbear remarking in this place, that the fact
obferved by Dr. *Darwin*, in the Philofophical Tranfactions
for 1788, fupports the theory we have here advanced, and
indeed, I think, cannot be fo rationally accounted for on any
other : the fact was, that air during its rarefaction attracts
heat

Evaporation from land in general muſt be leſs than the rain that falls upon land; otherwiſe there could be no rivers. In winter the evaporation is ſmall, compared to what it is in ſummer. From a ſeries of experiments made in the preſent year, 1793, I found the mean daily quantity evaporated from a veſſel of water, in a ſituation pretty much expoſed to wind and ſun, for 13 days of March, to be .033 of an inch in depth, the greateſt .064; for 21 days of April the mean daily quantity was .0555 of an inch, the greateſt .1115; for 26 days of May the mean was .0755, the greateſt .1346; for 14 days of June the mean was .063, the greateſt .098; for 8 days of July the mean was .122, the greateſt .195: I never found the evaporation from water any ſummer much to exceed .2 of an inch in 24 hours, in the hotteſt weather. From theſe experiments, and other conſidera-

T tions,

heat from the ſurrounding bodies, and gives off heat during its condenſation; now, the moment any quantity of atmoſpheric air is rarefied, its vapour muſt be rarefied alſo, and hence a portion of moiſture will expand into vapour in order to reſtore that ſtate of denſity which the temperature admits of, and abſorb the requiſite quantity of heat from the bodies adjacent; again, the moment air is condenſed, its vapour is condenſed proportionally, ſo that the abſolute quantity of vapour which retains its form, will always be as the *ſpace* occupied by the condenſed air, and the reſt will be precipitated, giving off its heat to the ſurrounding bodies.—Notwithſtanding what is here ſaid, it is probable that a decreaſed preſſure upon the ſurface of water *accelerates*, if it do not increaſe the evaporation, all other circumſtances being the ſame,

tions, it feems probable, that the evaporation
both from land and water, in the temperate and
frigid zones, is not equal to the rain that falls
there, even in fummer.

When a precipitation (or condenfation, which
ever it be) of vapour takes place, if the temperature of the air be above 32°, the matter precipitated is liquid, or in form of *rain;* but if the
temperature of the air be lefs than 32°, it is in
form of *fnow;* when drops of rain, in falling,
pafs throw a *ftratum* of air below 32°, they are
congealed, and form *hail.*

If we adopt the opinion, which to me appears
the more probable, that water evaporated is not
chymically combined with the aerial fluids, but
exifts as a peculiar fluid diffufed amongft the
reft; whenever any condenfation of it happens,
the matter muft be *precipitated*, though not in
the chymical fenfe of the word; we would therefore be underftood in this effay to ufe the words
precipitation and *precipitated* merely to denote the
effect, without any allufion to chymical agency.

Different theories to account for thefe preci-
pitations from the atmofphere have been formed;
but the principles of none appear to me to be
more plaufible, and confiftent with facts, than
that which has lately been offered to the public, in the *Edinburgh Philofophical Tranfactions*,
by

by Dr. *Hutton* of that place. From a fhort review of the article (for I have not feen the original) it appears, that he confiders the varieties
of heat and cold, affecting the folvent power of
the atmofphere, as the fole caufes of rain. Indeed, when we confider that evaporation and
the precipitation of vapour are diametrically oppofite, it is reafonable to fuppofe that they fhould
be promoted by oppofite caufes; and as heat
and dry air are favourable to evaporation, fo
cold, operating upon air replete with vapour,
promotes its precipitation. The point upon
which we differ, I fuppofe will be, that he confiders water chymically combined with the atmofphere, and that cold produces a precipitation
in a manner fimilar to what it does in water faturated with falt, or in other chymical proceffes;
whereas I fuppofe, that a portion of the vapour,
confidered as a diftinct and peculiar fluid, is
condenfed into water by cold; the effects refulting from the two theories will therefore be
much the fame.

The reafon then that a SW. wind in thefe
parts brings rain, feems to be, that, coming
from the torrid zone, it is charged with vapour,
and the heat efcaping as it proceeds northward,
a precipitation of the vapour enfues; but a NE.
wind, blowing from a cold into a warmer country,
has its capacity for vapour increafed, and therefore we generally find it promote evaporation.

T 2

From

From the obfervations upon the quantity of rain that falls in different places, it feems clearly afcertained, that there is more rain in mountainous than in level countries. The reafon feems to be, that the inferior, warm, and vapoury *ftrata* of air, ftriking againft the mountains, are made to afcend into the colder regions, by which means the vapour is precipitated : the fituation of places, however, may be too high to experience an extreme in this refpect; thus, the rain in *Switzerland*, and amongft the *Alps*, is not probably greater than in the north of *England*. It is more than probable too, that the rain in places fituate near the weftern coaft of *Great-Britain*, and of the Continent, is greater than in the more inland parts. Mr. *Clark*, in his Letters on the Spanifh Nation, obferves, that there was an inftance when no rain fell in *Caftile* for 19 months together; the province is in the centre of *Spain*, and at a great diftance from the fea.

In the level parts of this kingdom, and in the neighbourhood of the metropolis, the mean annual rain is only 19 or 20 inches.

Profeffor *Muffchenbroek* has given us an account of the mean annual rain at feveral places, which we fhall fubjoin, together with an account from fome other places. The inches differ a little in different countries, but the difference is too trivial to merit much notice in this place.

Mean

	Mean annual rain. Inches.
Utrecht, Haerlem, and Lifle, each -	24
Delf, and Harderwick, each - - -	27
Dort - - - - - -	40
Middleburgh, in Zealand - - -	33
Paris - - - - - -	20
Lyons - - - - - -	37
Rome - - - - - -	20
Padua - - - - - -	37½
Pifa - - - - - -	34¼
Zurick, in Switzerland - - - -	32
Ulm, in Germany - - - -	26⅝
Wittenberg - - - - -	16⅕
Berlin - - - - - -	19⅕
In Lancafhire - - - -	41
Upminfter, in Effex - - - -	19½

Bradford, in New England (2 years) * -	31.4
Langholm, } in Scotland† - -	36 +
Branxholm,}	31 +
Kendal - - - - - -	64.5
Kefwick - - - - -	68.5

From the table of the mean monthly rain at
Kendal and *Kefwick*, page 38, it appears, that if
we would pitch upon 6 fucceffive months, which
together produce more rain than any other 6
fucceffive months, at thefe places, we muft begin with September. At *Kendal*, from September to March there is 37.6 inches of rain, and
from March to September only 26.9 inches;
at *Kefwick*, the rain in the former period a-
mounts

* American Philofophical Tranfactions.

† Edinburgh Philofophical Tranfactions.

mounts to 40.4, and in the latter to 28.1.—
The reafon of this feems to be, that, in the former period, the temperature of the air is de-
creafing, and confequently its capacity for vapour
alfo; which circumftance is an additional caufe
of the precipitation of vapour. In the latter
period, the capacity of the air for vapour is increafing, which occafions a lefs precipitation.

When a precipitation of vapour takes place,
a multitude of exceedingly fmall drops form a
cloud, mift, or fog; thefe drops, though 800
times denfer than the air, at firft defcend very
flowly, owing to the refiftance of the air, which
produces a greater effect as the drops are fmaller,
as may be proved thus :— Let $d =$ the diameter
of a fmall drop, and $nd =$ that of a larger ; then
the refiftances, being as the fquares of the dia-
meters when the velocity is given, will be as d^2
and n^2d^2, refpectively ; but the magnitudes are
as d^3 to n^3d^3, or as 1 to n^3, whence, if the large
drop be divided into others of the fame magnitude as the fmall one, the number will be $= n^3$,
and the refiftance to them falling, as n^3d^2, whilft
the refiftance to an equal mafs in one drop is as
n^2d^2; confequently, the refiftance to the large
drop is to the refiftance of all the fmall ones,
moving with the fame velocity, as the diameter
of one fmall drop is to the diameter of the large
one, and the force being conftant, the time of
falling through a given fpace will be greater
                                          when

when the drops are fmall than when large. From this it appears, that clouds confifting of very fmall drops may defcend very flowly, which is agreeable to obfervation; if the drops in falling enter into a *ftratum* of air capable of imbibing vapour, they may be rediffolved, and the clouds not defcend at all; and if the air's capacity for vapour increafe, they may be all imbibed, and the cloud entirely vanifh. On the other hand, if the precipitation go forward, and the air below have its full quantity of vapour, the fmall drops meeting one another, will coalefce, and form larger ones, and defcend in form of rain to the earth's furface.—What is faid of rain, will likewife hold of fnow, except that the fmall particles coalefcing form flakes, by reafon of their not being fluid*.

From the important obfervations on the height of the clouds (page 41) we learn, that they are feldomer above the fummit of *Skiddaw*, in Nov. Dec. Jan. and Feb. than in the other months; this clearly indicates the effect of cold in reftraining the afcent of vapour. Were the meafurement extended above the fummit of the mountain, it is probable, from the apparent law of the table, that there could not be many obfervations

* This account of the nature of clouds, and of the mode of their rifing and falling in the atmofphere, was fuggefted by a philofophical friend and acquaintance; and it appeare to me very rational and confiftent.

fervations above 1300 yards in winter, nor above 2000 yards in fummer. This, it muſt be ob-ferved, relates to the height of the *under* ſurface of the groſs clouds only. The ſmall white ſtreaks of condenſed vapour which appear on the face of the ſky in ſerene weather, I have, by ſeveral careful obſervations, found to be from 3 to 5 miles above the earth's ſurface.

When vapour is condenſed into ſmall drops upon the ſurfaces of bodies on the ground, it is called *dew;* the only ſeeming difference betwixt dew and rain is, that the condenſation of the vapour in the one caſe is made at or near the ſurface of the body receiving it, and in the other the drops fall a confiderable ſpace before they reach the earth; the cauſe is the ſame in both caſes, namely, cold, operating upon vapoury air. At firſt view it will ſeem inconſiſtent that a con-denſation of vapour ſhould take place in the air reſting upon the earth's ſurface, which is gene-rally ſuppoſed to be warmer than that above; but it is an inconteſtable fact, that after ſun-ſet, and during the night, in ſerene weather, the air is coldeſt at the earth's ſurface, and grows warmer the higher we aſcend, till a certain moderate height (perhaps from 20 to 100 yards, or up-wards), this I have often obſerved myſelf, before I happened to ſee it elucidated, by a ſeries of experiments, in the *Lettres phyſiques, &c.* Tom. 5, page 561. And accordingly we find, that

dew

dew and hoar froft are more copious in valleys
than in elevated fituations. That dew depends
upon this circumftance can hardly be doubted,
becaufe when clouds or winds prevent it, there
is little or no dew formed.

We fhould fcarcely be excufed, in concluding
this effay without calling the reader's attention
for a moment to the beneficent and wife laws
eftablifhed by the Author of Nature, to provide
for the various exigencies of the fublunary crea-
tion, and to make the feveral parts dependent
upon each other, fo as to form one well regulated
fyftem, or whole.—In the torrid zone, and we may
add in the temperate and frigid zones alfo, in
fummer, the heat produced by the action of the
folar rays would be infupportable, were not a
large portion of it abforbed, in the procefs of
evaporation, into the atmofphere, without increafing its temperature; this heat is again given
out in winter, when the vapour is condenfed, and
mitigates the feverity of the cold. The dry
fpring months are favourable to agriculture, and
the evaporation, which then begins to be confiderable, abforbs a portion of the heat imparted
to the earth by the fun, and thus renders the
tranfition from cold to heat flow and gradual; in
autumn the fun's influence fails apace, and the
condenfation of vapour contributes to keep up
the temperature, and prevent too rapid a tranfition to winter.

V   ESSAY

ESSAY SEVENTH.

On the Relation betwixt the Barometer and Rain.

SINCE the barometer has become an inftrument of general ufe, and is adopted as a
guide by moft people interefted in the ftate of
the weather, it may be of fervice to invefti-
gate the relation fubfifting betwixt the weight
of the atmofphere and its difpofition for rain,
from the facts afforded us by obfervation,—and
we may at the fame time confider what further
arguments can be obtained in fupport of the
foregoing theories.

In the firft place it is remarkable, that, from
the table of the mean ftate of the barometer for
5 years, in page 16, we find the higheft mean
upon 6 fucceffive months obtained from March
to Auguft, inclufive; that is, the mean ftate of
the barometer for March, April, May, June,
July, and Auguft, taken together, is greater
than for any other 6 fucceffive months, being at
Kendal, for inftance, 29.83, and for the remaining 6 months, only 29.75. But what is more
particularly worthy of notice, is, that in this refpect, the rain and the barometer are juft the
reverfe of each other; for, in the former period
the

the rain was leaft, and greateft in the latter, as has been obferved, in page 141.

Again, by recurring to the tables, page 16 and 38, we fhall obtain the following arrangements of the months, beginning with that on which the mean ftate of the barometer was higheft, and proceeding regularly on to the loweft; and again, beginning with that month on which there was leaft rain, and proceeding to that on which there was moft.

Barometer high.	*Barometer low.*
May, Aug. June, Màr. Sept. April.	Nov. Feb. Octo. July, Dec. Jan.

Dry months.	*Wet months.*
Mar. June, May, Aug. April, Nov.	Octo. Feb. July, Sept. Jan. Dec.*

Now it is obfervable, that the evaporation is greateft from March to Auguft; confequently, the air is then farther from the point of faturation, or has a greater capacity for vapour, than in the other period; or, in other words, it is drier, relative to its temperature, than in the other period.—Hence then we have a ftrong ar-

V 2   gument

---

* By making the arrangements for *Kendal* alone, and taking in the prefent year, 1793, till Auguft, and part of 1787, we obtain the following:

Bar. May, Aug. June, Mar. Sep. April, Nov. Feb. July, Oct. Jan. Dec.
Rain. Mar. May, June, April, Aug. Oct. Nov. Feb. Sep. July, Jan. Dec.

gument for the theory of the barometer, as well
as for that of rain.

But to be more particular in the inveſtigation:
—It will be ſeen that there have been 6 months
when the mean ſtate of the barometer at *Kendal*
was 30 inches or above; 9 months when it was
29.9, or from thence to 30 inches; 17 months
when it was 29.8, or from thence to 29.9, &c.
as per the following table.—Now, in order to
examine the relation of the barometer and rain,
it will be proper to find the mean monthly rain
for thoſe diſtributions of the months when the
mean ſtate of the barometer was nearly the ſame.
This we have done, and the reſult follows.

Mean ſtate of the barometer, at Kendal.	Number of months.	Mean monthly rain in the different diſtributions, in inches.		
		Kendal.	Keſwick.	London*.
30 +	6	2.605	2.511	.212
29.9 +	9	3.362	4.018	.835
29.8 +	17	5.402	5.676	1.846
29.7 +	13	6.184	6.449	2.100
29.6 +	7	7.116	7.198	1.340
29.5 +	6	6.798	7.533	.898
29.4 +	1	3.306	3.600†	3.253
29.3 +	1	8.369	11.357	

The

* The account in this column, is the reſult of the 3 years' obſervations
we have inſerted in the firſt part; the firſt mean is for 4 months, when
the barometer at *London* was 30.1 plus; the ſecond for 6 months, when it
was 30 plus, &c. the reſt are for 7, 11, 5, 2, and 1 months, reſpectively.

† There was no rain-gauge this month at *Keſwick*; the quantity ſet
down is got by compariſon only.

The inferences to be drawn from this table are, 1ft, The higher the barometer is above its mean annual ftate, the lefs rain there is. 2d. The farther it is below its mean annual ftate, the more rain there is, till it comes to a certain point, after which the rain feems to decreafe again.

The firft of thefe inferences, being conformable to common obfervation, was expected; but the conclufion in the fecond, that the monthly mean ftate of the barometer may be *too low* to be attended with the *maximum* of rain, was not apprehended till the preceding table, which feems to warrant it, was digefted. However, it was immediately perceived, that the point might be cleared up, by felecting all thofe days which have produced the greateft quantity of rain, and finding the mean ftate of the barometer upon thofe days, which may be taken for that ftate moft conducive to the greateft quantity of rain. —The refult of a careful examination of my own obfervations, at *Kendal,* follows : during the extraordinary fall of rain on the 22d of April, 1792, (fee page 38) the mean of the barometer was 29.62; the other 2 days that gave more than 2 inches of rain each, the barometer was 29.59 and 29.33 refpectively : as for the other 56 days, on each of which there was more than 1 inch of rain, the mean ftate of the barometer upon the whole of them was 29.47, and for 54
of

of thofe days the barometer was between 29.03 and 29.81 ; the barometer on the other 2 days was plainly irregular, being on the one 28.5, and it is remarkable, that the rain of that day was barely 1 inch ; on the other it was 30.06, attended with an extraordinary circumftance. (See page 44, upon June 4, 1791).

From this it appears, that the heavieft rains may be expected when the barometer is about 29.47, at this place, or, in round numbers, 29½ inches, which is a little *above* the mean of the two great extremes obferved in January 1789, or 29.44.

In the laft 5 years there have been 1827 days, of which 1082, as per account, had rain, more or lefs, at *Kendal,* and 59 of thofe gave above 1 inch of rain each ; hence, at an average, there has been 1 of fuch days in every 31, wet and fair, and in every 18 wet days, nearly. The number of days when the mean ftate of the barometer was below 29 inches, were 40, of which 2 only were fair ; and yet there was but 1 of thofe that gave 1 inch of rain.   From thefe facts we may conclude, that when the barometer is very low, the probability of its being fair is much fmaller than at other times ; but that, on the other hand, the probability of very much rain, in 24 hours, is not fo great as at other times, which is confiftent with the conclufion
obtained

obtained from the facts ftated in the preceding paragraphs.

Upon an enumeration it appears, that there have been 78 days in the different months of the laft 5 years when the mean ftate of the barometer, at *Kendal,* was above the ufual high extreme for the month, as ftated at page 16; only 7 of thofe days were wet, and the rain in very fmall quantities; hence, the probability of a fair day at that place, to that of a wet one, in fuch circumftances, is as 10 to 1.

The preceding facts offer nothing but what appears confiftent with the theories of the barometer and rain; when the barometer is above the mean high extreme for the feafon of the year, the air muft, relatively fpeaking, be extremely *dry* or *cold,* or both, for the feafon; if it be extremely dry, it is in a ftate for imbibing vapour, and if it be extremely cold, no further degree of cold can then be expected, and therefore in neither cafe can there be any confiderable precipitation: on the contrary, when the barometer is very low for the feafon, the air muft relatively be extremely *warm* or extremely *moift,* or both; if it be extremely warm, it is in a fimilar ftate to dry air for imbibing vapour, and if it be extremely moift, there muft be a degree of cold introduced to precipitate the vapour, which cold, at the fame time, raifes
the

the barometer. From which it follows, that no very heavy and continued rains can be expected to happen whilft the barometer actually remains about the low extreme, but they muft rather be the confequence of a junction or meeting of extremes, which at the fame time effects a mean ftate of the barometer.

ESSAY

ESSAY EIGHTH.

On the Aurora Borealis.

AS this effay contains an original difcovery, which feems to open a new field of enquiry in philofophy, or rather, perhaps, to extend the bounds of one that has been, as yet, but juft opened; it may not perhaps be unacceptable to many readers to ftate briefly the train of circumftances which led the author to the important conclufions contained in the following pages.

It will appear, from the obfervations, that the author has been pretty affiduous, during the laft 6 years in noticing thofe very fingular and ftriking phenomena, the *auroræ boreales*, as often as they occurred; in which time he has alfo feen and confidered, with a proper attention, feveral conjectures and hypothefes, endeavouring to account for them; but as no hypothefis has yet appeared that explains the general phenomena in fuch a manner as to procure the acquiefcence of any rational enquirer, it was natural to expect that his attention would occafionally be turned towards an inveftigation of the nature and caufe of the *aurora*.

X   It

It feemed to be fufficiently proved that the *aurora* was not without the earth's atmofphere, though he had never feen any thing done which afcertained the real height of any one appearance with a tolerable degree of accuracy; and as the atmofphere, or at leaft the grofs part of it, is in all probability confined to the height of 15 or 20 Englifh miles, he was unwilling to admit of the greater height of the *aurora,* unlefs compelled to it by the refult of careful and accurate obfervations. The prevailing idea too that the *aurora* may be *heard,* was another means to induce him to think it was at a moderate height. —Appearances, however, were in direct oppofition to the thought;—that one and the fame *aurora* fhould be feen over a vaft extent of country, with much the fame circumftances, and that fome of them fhould appear in *France, Spain,* and *Italy,* whilft they fo very feldom pafs our zenith in the north of *England,* was a very ftrong argument for their great height. The beft obfervations likewife upon thofe large fiery meteors which occafionally fly over the country, and are feen at fuch diftant places, feem to prove the exiftence of an elaftic fluid at the height of 60 or 80 miles at leaft, which far exceeds the height of the atmofphere as prefcribed by the obfervations upon the barometer, or even by the twilight; and if the atmofphere exceed the height of 45 or 50 miles, as determined by the obfervations on the duration of

twilight,

twilight, we have no *data* from whence to fix its
bounds; it may, for ought we know, amount to
4 or 5 hundred miles.

These confiderations, it is evident, could not
fail of fuggefting to the author the expediency
of determining, by actual obfervations, the real
height of the *aurora borealis*. This he thought
might be accomplifhed by the affiftance of his
friend and colleague in the bufinefs, Mr. *Crofthwaite*, of *Kefwick*, who having for a long time
been accuftomed to make fuch obfervations, was
the more eligible for the purpofe; but the manner of doing it was firft to be determined upon,
as the great difficulty was to afcertain that the
obfervations were cotemporary, and made upon
one and the fame object.

As the *aurora* often confifts of upright beams,
efpecially when high above the horizon, and thefe
feldom continue one minute the fame, the queftion
was, whether to attempt the altitude of the bafe of
the beams, or the vertex, or both; this put the
author upon confidering more particularly what
the real form of the beams is when ftript of the
optical illufion, which muft accompany all objects feen at a great diftance in the atmofphere,
namely, that of appearing to coincide with the
blue vault, or fky, and to conftitute a part of its
fpherical furface. A very moderate fkill in optics was fufficient to convince him, that as the

X 2   luminous

luminous beams at all places appear to tend towards one point about the zenith, they muſt in
reality be ſtraight beams, parallel* to each other,
and nearly perpendicular to the horizon ; and
from the appearance of their breadth, they muſt
be cylindrical. Theſe circumſtances accounted
at once for the *aurora* appearing ſo denſe northward, towards the horizon, and the beams being
ſo thin and ſcattered towards the zenith, which
is ſo uniformly the caſe.   Moreover, as the
beams appear to riſe above each other in regular
ſucceſſion one ſet above another, in ſuch ſort,
that the higher the baſes of the beams are, the
higher are their vertices, it ſeemed from this
circumſtance probable, that they are all of the
ſame length and height; if this be the caſe, by
determining the greateſt angle ſubtended by the
beams, the relation or proportion of their length
to their height above the earth's ſurface may be
determined geometrically. —This circumſtance
deſerved to be kept in view; and it appeared,
from obſervations made upon the *aurora* afterwards, that though the fact could not eaſily be
aſcertained, yet ſo much was certain, that the
length of the beams bore a very great proportion to their diſtance from the earth, even ſo as
to equal or perhaps ſurpaſs the ſaid diſtance.

Thus

* The author did not ſee, before May 1793, the Philoſophical Tranſactions for 1790, in which he finds this idea is
ſuggeſted by *H. Cavendiſh*, Eſq. F. R. S. and A. S.

Thus ftood the author's knowledge and ideas upon the fubject in the autumn of 1792.—The very grand *aurora* in the evening of the 13th of October, was that which firft fuggefted and led to the difcovery of the relation betwixt the phenomenon and the earth's magnetifm. When the theodolite was adjufted without doors, and the needle at reft, it was next to impoffible not to notice the exactitude with which the needle pointed to the middle of the northern concentric arches : foon after, the grand dome being formed, it was divided fo evidently into two fimilar parts, by the plane of the magnetic meridian, that the circumftances feemed extremely improbable to be fortuitous ; and a line drawn to the vertex of the dome, being in direction of the *dipping-needle*, it followed, from what had been done before, *that the luminous beams at that time were all parallel to the dipping needle.* It was eafily and readily recollected at the fame time, that former appearances had been fimilar to the prefent in this refpect, that the beams to the eaft and weft had always appeared to decline confiderably from the perpendicular towards the fouth, whilft thofe to the north and fouth pointed directly upwards, the inference therefore was unavoidable, that the beams were guided, not by gravity, but by the earth's *magnetifm,* and the difturbance of the needle that had been heretofore obferved during the time of an *aurora*, feemed to put the conclufion paft doubt. It

was

was proper however to obferve whether future appearances correfponded thereto, and this has been found invariably the cafe, as related in the obfervations.

Soon after this, the author wrote to Mr. *Crofthwaite*, defiring him to pay particular attention to thefe phenomena for a feafon, to take the bearings, altitudes, times, &c. of every remarkable appearance, and to obferve the point to which the beams converged, the bearing of the perpendicular beams, the extent and bearing of the large, northern, horizontal lights, &c. Thefe he performed with much readinefs and fkill, and his obfervations agree fufficiently with thofe made at *Kendal*, though he was entirely unacquainted with the difcovery, and confequently his obfervations could not be warped to fuit the author's purpofe.

The obfervations on the 15th of February, 1793, are thofe upon which the height of the *aurora* refts principally, as none of the others were fufficiently well timed and circumftanced to be fubfervient to this purpofe, except perhaps that on the 30th of March, 1793. *

We

* It may not be improper here to advert to a circumftance, which, if not noticed, may be a means of fubjecting the author, in fome degree, to the imputation of plagiarifm. ——The advertifements refpecting this work were printed on the 10th of April, 1793, in which the difcovery above mentioned

We fhall now proceed to ftate the different parts of this effay, difpofing them into feparate fections, as follows.

SECTION

mentioned was announced as an original one, and never before publifhed; the author not knowing that any one had publifhed the moft diftant intimation of their afcribing the phenomena of the *aurora borealis* to magnetifm. On the 17th of faid month, *George Birkbeck*, of *Settle*, an ingenious and intelligent young man, a fubfcriber to this work, informed the author, that an anonymous perfon, in a certain periodical publication, had given an effay on the *aurora borealis*, in which, amongft other conjectures, he had advanced the opinion that it might be occafioned by the earth's magnetifm;—he was fo obliging as to tranfmit the author a copy of the effay itfelf, which may be feen in a work entitled *Mathematical, Geometrical, and Philofophical Delights*, No. 1." publifhed May 1, 1792, under the infpection of a Mr. *Whiting.*

The author, who fubfcribes himfelf *Amanuenfis*, ftates his conjectures to the following purport, viz.

1ft. He fuppofes that magnetic effluvia are conftantly iffuing from the earth's magnetic pole in the north, and that thefe effluvia, which he confiders of a ferruginous nature, fly off in every direction along the magnetic meridians; he then conjectures that the fulphurous vapours, rifing from the many volcanos in the north, mixing with the magnetic effluvia, may catch fire, and fulgurate.

2d. He conjectures that inflammable air having caught fire, may receive a magnetic direction, by the current of magnetic effluvia; he fubjoins to this conjecture, fome very juft obfervations on the *aurora*, which we fhall have occafion to mention hereafter.

3d. He conjectures that " a highly fubtilized aerial nitre " always enters into the compofition of an *aurora.*"

4th.

SECTION FIRST.

Mathematical Propoſitions neceſſary for illuſtrating and confirming thoſe concerning the
Aurora Borealis.

PROPOSITION I.

ALL lines or ſmall cylinders, whether ſtraight, curved, or crooked, ſeen at a conſiderable diſtance in the atmoſphere, and ſituate within a plane paſſing through the eye, muſt appear arches

4th. That the *aurora*, like lightning, may be of an electric nature.

5th. He aſks, " May the luminoſity be conveyed on the " magnetic effluvia, as the electric on an iron wire ?"

6th. He conceives the reaſon why the *aurora* is ſo frequent now is, becauſe there are more volcanos in the north.

I ſhould ſuppoſe that theſe conjectures, as far as they refer the phenomena of the *aurora borealis* to magnetiſm, are original; and from the time of the publication it might be ſuſpected that I received the firſt hint from it; this however was not the caſe, this work being nearly ready for the preſs before the 10th of April, and it was not till after, that the letter containing the eſſay came to hand, which firſt furniſhed me with the preceding conjectures; beſides, it will be ſeen that my opinions are, for the moſt part, very different from thoſe ſtated above.——It is not meant by this to depreciate the merit of the ingenious *Amanuenſis*, who will probably be well ſatisfied to ſee that the ſuppoſition of a relation between the *aurora borealis* and magnetiſm, which probably firſt occurred to him, is capable of being proved to a demonſtration.

arches of a circle, in whofe centre is the eye,
bounded by lines drawn from the eye to the extremities of the objects.

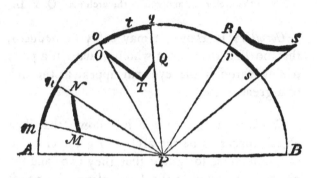

DEMONSTRATION.

The femicircle *A m n o t q r s B P* reprefents a part of
any plane paffing through the eye, fuppofed to be at *P*, the
centre; *A P B* the interfection of the faid plane with the
plane of the horizon; the arch of the femicircle reprefents
the interfection of the firft mentioned plane with the blue
canopy or fky; *MN, OTQ*, and *RS* reprefent three cylindrical beams feen at a diftance, whofe axes are in the
plane *A m n o t q r s B P* indefinitely extended. Then the
object *MN* being at a confiderable diftance, as 5, 10, &c.
miles, and quite detached from all objects on the earth's
furface, it follows, from the principles of optics, that the
mind cannot judge with certainty either of the abfolute diftance of the object, or whether the extremity *M* or *N* is
more diftant; in fuch a cafe, therefore, nothing appears to
the contrary but that both ends are equally diftant, and that
MN is an arch of a circle in the fky, with the eye in the
centre; and this in fact is the judgment that is uniformly
made in the cafe. For it is known to every one, that ce-
leftial objects. and objects at a diftance in the air, as the fun,

Y   moon,

moon, ſtars, meteors, &c. all *appear* at the ſame diſtance, though nothing can be more diſproportionate than their real diſtances; that is, they all appear as if ſituate in the ſky: hence then the objeƈt *MN* will appear as the arch *mn*, *OTℒ* as the arch *otq*, and *RS* as the arch *rs*. Q. E. D.

Corollary 1. Hence it may eaſily be deduced, that no line that is not wholly ſituate in a plane paſſing through the eye can appear as the arch of a great circle.

Corollary. 2. Hence alſo it follows, that if an objeƈt appear to be the arch of a great circle to two obſervers, ſo ſituate that they two, and the objeƈt, are not all in the ſame plane, the objeƈt muſt be a ſtraight line, or ſmall cylinder, becauſe it muſt neceſſarily be wholly in two planes, and conſequently in their common interſeƈtion, which is a ſtraight line *(Euclid,* 11 and 3*)*.

PROPOSITION II.

Imagine a cylindrical beam, as *AE*, elevated in the air, and viewed from a ſtation on the earth, at a diſtance, as in the laſt propoſition; and ſuppoſe the beam ſo ſituate that a perpendicular *CP* from *C* to the ſide of the cylinder *BE* may fall below *B*, or in the prolongation of *EB;* then, I ſay, the beam will apper broadeſt near the bottom, and narrower as it aſcends, that is, its ſides will appear bounded by the circumferences of two great circles, having their common interſeƈtion in a line *CV* parallel to *BE*.

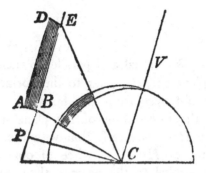

DEMONSTRATION.

By the laſt propoſition the lines bounding the cylinder longitudinally will appear as arches of great circles; and if the line BE be ſuppoſed to be extended indefinitely, the angle PCE increaſes, and when BE becomes infinite, CE coincides with CV, and the angle $PCV =$ a right one; and the very ſame concluſion will follow if a perpendicular be let fall from C upon AD, or any other line parallel to BE; therefore all right angles parallel to BE will appear arches of circles, which, if prolonged, would interſect each other in the line CV, and the ſpace bounded by any two arches will grow narrower from P towards V. Q. E. D.

Corollary. If there be a number of beams ranged all over a tranſparent plane parallel to the horizon, at the height of AB; and if theſe beams be parallel to the beam AE, then they will all appear to converge towards V, from every point of the horizon.

Scholium. The appearances of the extremities of the cylinder are not here conſidered; but it would be eaſy to prove they muſt appear elliptical.

PRO-

PROPOSITION III.

Let there be a feries of cylindrical beams, *DF, KL*, &c. equal and parallel to each other, all in a plane perpendicular to the horizon, and at equal diftances from the horizon; and let *AB* be the interfection of the plane with the horizon; *HvI* its interfection with the fky; *C* the centre of *HvI*, the place of obfervation; and *Cv* parallel to the beams; then, firft, the beams will appear to rife above each other fuc‑ceffively, in the fky, in fuch fort, that, of any two beams, that which has the higher bafe, will have the higher vertex alfo, except when the beams appear to pafs through, or lie wholly be‑yond the zenith; fecond, thefe about the zenith will appear broadeft, and thofe neareft the ho‑rizon narroweft.

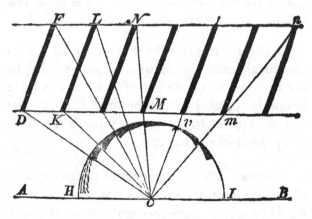

DEMONSTRATION.

Join *CD, CF, CK,* and *CL*; then the bafe *K* will appear
higher than the bafe *D* by the angle *DCK,* and the vertex
L higher than the vertex *F* by the angle *FCL,* and fo on
for the reft of the beams, till the angle reprefented by *FCL*
is equally divided by a line from *C* to the zenith; afterwards
the contrary takes place. The angle under which the diameter of the beam appears, being fuppofed fmall, will be
nearly as the diftance inverfely, and therefore greateft at
the zenith, and lefs below, in proportion as radius to the
fine of elevation. Q. E. D.

PROPOSITION IV.

The fame Figure remaining.

If the beams are equidiftant, and if *C M N,*
C m n be drawn on each fide of *v,* fo as to touch
the bafes of two beams in *M* and *m,* and the
vertices of the two next beams in *N* and *n ;*
then all the beams included in the angle *NCn*
will appear diftinct, and all thofe below, on both
fides, will partly cover each other, if opaque;
but if luminous, the light of the different beams
being blended, will increafe in denfity downward, according to the number of beams croffed
by a right line from *C.*

DEMONSTRATION.

The firft part is obvious, from the elements of geometry;
and from the principles of optics, the diftance of the beams
makes no difference in their apparent brightnefs, unlefs what
arifes from the want of perfect tranfparency in the atmofphere,

phere, which fomewhat obfcures diftant objeɛts; hence, the greater the number of beams croffed by a right line from *C*, the denfer will be the light in that direɛtion. Q. E. D.

PROPOSITION V.

The fame things being fuppofed as in Propofition third: let a circle be defcribed through the extremities of any one beam, as *D F*, to touch the horizontal line in *c**; and if *c D* and *c F* be joined, the angle *D c F*, fubtended by the beam, will be greater than that fubtended by any other beam, as feen from *c*; and if *F D* be produced to meet the horizon in *A*, and the quantity of the angle *D c F* be given, the proportion of *AD* to *DF* may be determined.

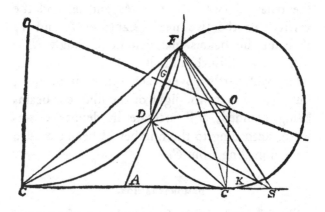

DEMON-

* To do which, fee the laft book of *Simpfon's Geometry,* prob. 42. third edit.

DEMONSTRATION.

Draw any line, DKS, to cut the circle in K, and meet the horizontal line in S; join FK and FS; then the angle $DKF =$ angle DcF *(Euclid,* 3, 21); and angle DKF is greater than DSF *(Euclid,* 1, 21).

Draw oc perpendicular to the horizon from the centre of the circle o, and bifect DF by the perpendicular oG, and join oD; then, fince the angle DcF is given, DoG is given alfo, being $= DcF$ *(Euclid,* 3, 20); alfo the angles G and Aco being right, and angle A given by hypothefis, angle Goc is given alfo, and confequently Doc; and the triangle Doc being ifoceles, the angles at D and c are both given, and angle AcD alfo, being the complement of Dco; whence it will be

Sine AcD : fide AD :: fine A : fide cD;

And fine Doc : fide cD :: fine Dco : fide Do;

And radius : fide Do :: fine DoG : fide $DG = \frac{1}{2}DF$, which gives the ratio of AD to DF. Q. E. D.

Scholium. We have here fuppofed the angle A acute; but if it be taken obtufe, or the obfervations be made on the other fide of A, the proportion of $AD : DF$ may be found equally, but the greateft angle under which the beams appear will be lefs; thus, if oG be produced to O, fo that upon O, as a centre, a circle may be defcribed to pafs through F and D, and touch the horizontal line in C; then, the greateft angle DCF will be at C, where the circle touches the horizontal line, as before.

SECTION

SECTION SECOND.

Phenomena of the Auroræ Boreales.

THE appearances of the *aurora* come under four different defcriptions.——Firft, a *horizontal light*, like the morning *aurora*, or break of day. —Second, fine, flender, luminous *beams*, well defined, and of denfe light; thefe continue $\frac{1}{4}$, $\frac{1}{2}$, or 1 whole minute, fometimes at reft apparently, but oftener with a quick lateral motion.——Third, *flafhes* pointing upward, or in the fame direction as the beams, which they always fucceed; thefe are only momentary, and have no lateral motion, but they are generally repeated many times in a minute; they appear much broader, more diffufe, and of a weaker light, than the beams; they grow gradually fainter till they difappear. Thefe fometimes continue for hours, flafhing at intervals.——Fourth, *arches*, nearly in the form of rainbows; thefe, when complete, go quite acrofs the heavens, from one point of the horizon to the oppofite point.

When an *aurora* takes place, thofe appearances feem to fucceed each other in the following order :—Firft, the faint rainbow-like arches; fecond, the beams; and, third, the flafhes: as for the northern horizontal light, it will appear in the fequel to confift of an abundance of

flafhes,

flaſhes, or *beams*, blended together, owing to the ſituation of the obſerver relative to them.

Theſe diſtinctions, and the terms appropriated for them, muſt be kept in view, in attending to the following phenomena.

PHENOMENON I.

The beams of the *aurora borealis* appear, at all places alike, to be arches of great circles of the ſphere, with the eye in the centre, and theſe arches if prolonged upwards would all meet in one point.

This is conformable to my own obſervations, and to all the accounts I have ſeen of the *aurora*.

PHENOMENON II.

The rainbow-like arches all croſs the magnetic meridian at right angles; when two or more appear at once, they are concentric, and tend to the magnetic eaſt and weſt; alſo, the broad arch of the *horizontal light* tends to the magnetic eaſt and weſt, and is biſected by the magnetic meridian; and when the *aurora* extends over any part of the hemiſphere, whether great or ſmall, the line ſeparating the illuminated part of the hemiſphere from the clear part, is half the circumference of a great circle, croſſing the magnetic meridian at right angles, and terminating

Z   in

in the magnetic eaſt and weſt; moreover, the beams perpendicular to the horizon are only thoſe on the magnetic meridian.

Theſe have been the uniform appearances at *Kendal* for a ſeries of obſervations paſt, as has been related; and from re-collection, and the notes made upon former appearances, as well as from the inference to be drawn from the later obſer-vations, I have no doubt the whole liſt of the *aurora* were conformable to this deſcription.

The accounts from *Keſwick* corroborate the ſame; the hori-zontal light is deſcribed as extending from WSW. to ENE. and its higheſt part in the middle, or NNW. or, when paſt the zenith, SSE *.——As for the vertical ſtreamers, their de-clination from the vertical circles being ſo ſmall, except about the eaſt and weſt points, it is no wonder if there be ſome la-titude in theſe obſervations, when the eye is to judge; we do not find, however, that this latitude has exceeded 10° from the magnetic meridian.

That this phenomenon agrees with the obſervations made in *England*, *France*, *Germany*, &c. in the beginning of this century, when the *aurora* fiſt appeared, we learn from the following extracts from the Tranſactions of the Pariſian Academy.

1707. March 6, between 7 and 10 in the evening, M. *Leibnitz* ſays an *aurora borealis* was obſerved at *Berlin;* there were two luminous arches, one above the other, both directly northward,

---

* The horizontal arches, indeed, do not always appear to extend juſt to the magnetic eaſt and weſt, but often to fall ſhort of, and ſometimes to ſurpaſs thoſe points; the reaſon is, we judge of its extent from its viſibility above the ſenſible horizon, and the light is either ſo faint by the great diſ-tance, or objects intervene, that we ſeldom ſee the extremity of the arch, within 2 or 3° of the horizon; this contracts or enlarges its viſible extent amaaingly, when the arch makes a ſmall angle with the horizon.

northward, their concavity turned downwards, their chords parallel to the horizon.——The variation of the needle in *Germany*, &c. at that time, was very little from the true north.

1716. M. *Miraldi* defcribes the horizontal lights of the 11th, 12th, and 13th of April, as having all the fame fituation, namely, extent from 45 or 50° W to 35 or 30° eaſt of the meridian.——The variation at *Paris* was then about 1 point weſterly.

——March 17. a rainbow-like arch was ſeen at *Breſt;* it extended from E. to W. croſſing the meridian ſouth of the zenith; ſoon after, a horizontal light was ſeen, extent from NW to NNE.

——At *Rouen*, the fame night, a horizontal light was ſeen; its extent from 10° E. to 25 or 30° W.

——At *Newark*, in *Nottinghamſhire*, it was ſeen between the NW. and N.

One ſeen at *Copenhagen*, February 1, 1707, is ſaid to have extended from WNW. to NNE.

September 12, 1621. *Gaſſendus* obſerved an horizontal light, at *Aix*, in *Provence;* it extended between the ſummer riſing and ſetting.——N. B. The variation then was a little to the eaſtward.

1718. March 4. M. *Miraldi* obſerved an horizontal light; extent from NW. to NE. but declining about 10° more to the weſt.

These obſervations, compared with thoſe recently made, ſufficiently indicate that the poſition of the *horizontal lights* and *arches*, changes with the needle, and is now much more weſterly than formerly*.

Z z   It

———————————————————

* Since writing the above, I find in the Philoſophical Tranſactions of the Royal Society for 1790, vol. 80. ſeveral accounts of the rainbow like arches. In Art 3. Mr. *Hey*, after defcribing ſeveral arches, ſays, "the poles of all the " complete arches which I have ſeen had a *weſtern* variation from the pole " of the equator"——In Art. 5. Mr. *Hutchinſon* defcribes one ſeen on the 23d of February, 1784, at *Kimbolton*, (63 miles NNW. of *London*) to have extended from ENE. to WSW; and a defcription of the ſame appearance, not differing eſſentially, is given in Art. 4.

It fhould, however, be obferved, that this phenomenon is to be underftood as *general,* rather than *univerfal;* becaufe the *horizontal lights,* and *arches,* are fometimes interrupted, which caufes the *aurora* to be feen occafionally almoft wholly to the eaft or weft of the magnetic meridian ; but on all fuch occafions I have obferved the inclination of the *beams* invariably the fame, in the fame quarter of the heavens, as far as the eye could judge.———In fact, if the *horizontal lights,* &c. were not interrupted, the zone of light muft quite furround the northern parts of the earth, at every appearance, which we are pretty certain is feldom, if ever, the cafe.

PHENOMENON III.

That point in the heavens to which the *beams* of the *aurora* appear to converge at any place, is the fame as that to which the fouth pole of the *dipping-needle* points at that place.

Granting the truth of the two preceding phenomena, it follows, that the point of convergency muft be in the magnetic meridian ; and this point, from the beft obfervations I can make, is between 70 and 75° from the fouth ; which agrees with the obfervations at *Kefwick :* and it appears that the *dipping-needle* in *England* points to that part.——My notes upon the *auroræ* for 4 or 5 years paft ftate the point of convergency to the fouth of the zenith, when a crown was formed, and I believe the remark has been generally made, wherever the appearance was feen and attended to.——*Kircher* obferved the point 29° fouth of the zenith, at *Berlin.*

In fupport of the two laft phenomena I might alfo quote the ingenious *Amanuenfis* whom I have mentioned in the introduction to this effay ; he fays, " that the lucid columns, " or radiating flafhes of the *aurora borealis* almoft always " fhoot off from the north to the fouth, correfponding in a
" great

" great meafure to the magnetic meridian. And I have
" conftantly obferved" (adds he) " the *corona*, concourfe,
" or concentration, if I may fo call it, of thefe lucid rays
" near the zenith, fo much to the eaft of it as anfwered
" nearly to the weftern declination of the common magnetic
" needle,—and I think I never obferved the *corona* to the
" weftward of it.

PHENOMENON IV.

The *beams* appear to rife above each other in
fucceffion, fo that of any two beams that which
has the higher bafe has the higher fummit alfo,
or its fummit nearer the point of concourfe;
the angle fubtended by the length of each beam
is not the fame, it being greateft about half way
from the horizon to the zenith, and lefs above
and below; alfo the beams to the fouth fubtend
lefs angles than thofe to the north, having the
fame altitude.—The greateft angle to the north
feems to be about 25 or 30°; and that to the
fouth 15 or 20°.

PHENOMENON V.

Every *beam* appears broadeft at or near the
bafe or bottom, and to grow narrower as it
afcends, in fuch fort that the continuation of its
bounding lines would meet in the common centre to which the beams tend; yet the fummit of
the beam is not flat, but pointed.—The higheft
beams feem about 3° broad, and the loweft 1°.

The

The two laft phenomena are the refult of my own obfervations chiefly; but there is fome difficulty and uncertainty
in meafuring the angles fubtended by the lower beams, by
reafon of their being one behind another; it muft therefore
be left to future obfervations to determine more accurately
the angles under which the beams appear in different parts
of the hemifphere.

SECTION THIRD.

Propofitions concerning the Aurora Borealis.

PROPOSITION I.

THE luminous *beams* of the *aurora borealis,*
are cylindrical, and parallel to each other, at
leaft over a moderate extent of country.

The beams muft be parallel to each other, from Corol. to
Prop. 2, and Corol. 2, Prop. 1, Sect. 1; and from Phenom.
1. Hence, and from Prop. 2, Sect. 1, and Phenom. 5,
they are cylindrical.

PROPOSITION II.

The cylindrical beams of the *aurora borealis*
are all *magnetic,* and parallel to the *dipping-needle*
at the places over which they appear.

From the Corol. to Prop. 2, Sect. 1, and Phenom. 3, it
follows, that the beams are parallel to the *dipping-needle;*
and as the beams are fwimming in a fluid of equal denfity
with themfelves, they are in the fame predicament as a magnetic bar, or needle, fwimming in a fluid of the fame fpecific
gravity

gravity with itfelf; but this laft will only reft in *equilibrio* when in the direction of the *dipping-needle*, owing to what is called the *earth's magnetifm*; and as the former alfo refts in that pofition only, the effects being fimilar, we muft, by the rules of philofophizing, afcribe them to the fame caufe.— Hence, then it follows, that THE AURORA BOREALIS IS A MAGNETIC PHENOMENON, AND ITS BEAMS ARE GOVERNED BY THE EARTH'S MAGNETISM *.

PROPOSITION III.

The height of the *rainbow-like arches* of the *aurora*, above the earth's furface, is about 150 Englifh miles.

This appears from the calculation made from the obfervations on the 15th of February, 1793,—but other obfervations ought to be made at more diftant places, to afcertain the height with more precifion. Poffibly the height may be different at different times†.

PRO-

* I am aware that an objection may be ftated to this;---If the beams be fwimming in a fluid of equal denfity, it will be faid they ought to be drawn down by the action of the earth's magnetifm. Upon this I may obferve, that it is not my bufinefs to fhew why this is not the cafe, becaufe I propofe the magnetifm of the beams as a thing demonftrable, and not as an hypothefis. We are not to deny the caufe of gravity, becaufe we cannot fhew how the effect is produced.---May not the difficulty be leffened by fuppofing the beams of *lefs* denfity than the furrounding fluid.

† Since writing the above, I find Mr. *Cavendifh* has, in Art. 10 of the Philofophical Tranfactions for 1790, calculated the height of an arch obferved at different places, on the 23d of February, 1784, to be betwixt 52 and 71 miles.---But, with deference, I would remark, that the obfervations above mentioned appear to me better circumftanced than thofe upon which his calculation is founded, and therefore the refult of them more to be relied upon.

PROPOSITION IV.

The beams of the *aurora* are fimilar and equal in their real dimenfions to one another.

This is not capable of ftrict demonftration, for want of more exact obfervations; it is, however, rendered extremely probable from Prop. 3 and 5, Sect. 1, and Phenom. 4 and 5. —Indeed the phenomena are almoft irreconcileable to any other fuppofition.

PROPOSITION V.

The diftance of the *beams* of the *aurora* from the earth's furface, is equal to the length of the beams, nearly.

Allowing the truth of the laft propofition, and comparing Prop. 5, Sect. 1, with Phenom. 4, we fhall find the pheno-menon to agree beft with the fuppofition of the equality of the diftance and length of the beams.

We have here fubjoined the refult of a calculation of the angles fubtended by the beams, on three different fuppofi-tions, namely, 1ft, when the length of the beams is equal to their diftance from the earth; 2d, when the length is but half that diftance; and, 3d, when it is twice the diftance.— The calculation is eafily made by inverting Prop. 5, Sect. 1, and fuppofing the point c variable, where we have the ratio of AD to DF, inftead of the angle DcF given; the beams are fuppofed to be thofe in the plane of the magnetic meri-dian, both north and fouth of the zenith, and their bafes are taken at 10°, 20°, 30°, &c. altitude. The angle FAc is fuppofed 72°.

When accurate obfervations fhall be made, I have no doubt the angles on the 2d fuppofition will be found too little, and thofe on the 3d too great.　　　　　**Angles**

Angles AcD & ACD.	AD : DF :: 1 : 1 Angle DcF.	Angle DCF.	AD : DF :: 1 : ½ Angle DcF.	Angle DCF.	AD : DF :: 1 · 2 Angle DcF.	Angle DCF.
10°	10° 30	8° 27	5° 14	3° 25	20° 51	15° 23
20	19 32	13 4	10 7	7 16	35 2	21 27
30	24 52	14 12	13 42	8 22	40 10	21 33
40	26 34	12 50	15 33	7 56	39 46	18 27
50	25 36	9 48	15 43	6 16	36 27	13 36
60	22 48	5 43	14 32	3 45	31 23	7 45
70	18 53	1	12 21	0 40	25 26	1 20
80	14 15		9 28		18 58	
90	9 14		6 11		12 18	
100	4 4		2 44		5 24	

Scholium. It is very probable the *rainbow like arches* are either at the top or bottom of the *beams*, and I am inclined to think they are at the top, not only becaufe their light is faint, but becaufe the beams fhould be feen at a much greater diftance than it feems they are, if they were 300 miles high, or twice the height of the arches; and the obfervations on the 30th of March, 1793, feem to confirm the opinion of the bafes of the beams being 60 or 70 miles high, or about half the height of the arches.

If the fummits of the beams be 150 miles high, their bafes will, according to this propofition, be 75 miles high, and the whole length of the beams about 75 miles, or, more nearly, 75 miles × $\frac{\text{radius}}{\text{fine of } 72°}$. And if the diameter of the bafe be $\frac{1}{10}$ of the length, each luminous beam will be a cylinder of $7\frac{1}{2}$ miles in diameter, and 75 miles long*.

A a   N. B.

* If a magnet be required to be made of a given quantity of fteel, it is found by experience to anfwer beft when the length is to the breadth as 10 to 1 nearly : it is a remarkable circumftance that the length and breadth of the magnetic beams of the *aurora* fhould be fo nearly in that ratio.--- Query, if a fluid mafs of magnetic matter, whether elaftic or inelaftic, were fwimming in another fluid of equal denfity, and acted on by another magnet at a oiftance, what form would the magnetic matter affume? Is it not probable it would be that of a cylinder, of proportional dimenfions to the beams of the *aurora* ?

N. B. An object elevated 75 Englifh miles may be feen at the diftance of 10 geographical degrees; if elevated 150 miles, it may be feen 14°; if 300 miles, 20°.

PROPOSITION VI.

That appearance which we have called the *horizontal light*, and which is always fituate near the horizon, is nothing but the blended lights of a group of *beams*, or *flafhes*, which makes the appearance of a large luminous zone.

The figure to Prop. 3, Sect. 1, reprefents a feries of beams fuch as thofe of the *aurora*, fituate in the plane of the magnetic meridian, and *C* the place of obfervation. And it is proved in Prop. 4, Sect. 1, that the lights of the diftant beams in that plane will be blended, to a certain elevation, to the obferver at *C*. Imagine a feries of planes parallel to the plane of the magnetic meridian, with beams fituate in them likewife; then, from the principles of optics, the rows of beams in every two of the planes will appear to approach each other, as the diftance from the obferver increafes; and when that diftance becomes indefinitely great they will all feem to coincide; hence the beams will appear blended, both horizontally and perpendicularly, and will confequently conftitute a large zone of denfe light. This zone muft appear at right angles to the magnetic meridian, becaufe it is obferved (Phenom. 2.) that when the beams of the *aurora* extend over a great part of the hemifphere, they are always bounded by an arch croffing the magnetic meridian at right angles.

SECTION

SECTION FOURTH.

Theory of the Aurora Borealis.

IN the preceding fection we have deduced the nature of the *aurora,* merely by combining mathematical principles with the phenomena ; the conclufions, therefore, are not drawn from *hypothefis,* but from *facts,* and muft hold, as far as the facts are well afcertained, and the principles truly applied.—In this fection we mean to propofe fomething by way of hypothefis, to account for thofe phenomena.

The *light* of the *aurora* has been accounted for on three or more different fuppofitions : 1. It has been fuppofed to be a flame arifing from a chymical effervefcence of combuftible exhalations from the earth. 2. It has been fuppofed to be inflammable air, fired by electricity. 3. It has been fuppofed electric light itfelf.

The firft of thefe fuppofitions I pafs by, as utterly inadequate to account for the phenomena. The fecond is preffed with a great difficulty how to account for the exiftence of *ftrata* of inflammable air in the atmofphere, fince it appears that the different elaftic fluids, intimately mix with each other ; and even admitting the exiftence of thefe *ftrata,* it feems unneceffary to in-

A a 2   troduce

troduce them in the cafe, becaufe we know that
difcharges of the electric fluid in the atmofphere
do exhibit light, from the phenomenon of lightning.—From thefe, and other reafons, fome of
which fhall be mentioned hereafter, I confider it
almoft beyond doubt that the *light* of the *aurora
borealis*, as well as that of *falling ftars* and the
larger meteors, is electric light folely, and that
there is nothing of combuftion in any of thefe
phenomena.

Air, and all elaftic fluids, are reckoned a-
mongft the non-conductors of electricity. There
feems, however, a difference amongft them in
this refpect; dry air is known to conduct worfe
than moift air, or air faturated with vapour.
Thunder ufually takes place in fummer, and at
fuch times as the air is highly charged with vapour; when it happens in winter, the barometer
is low, and confequently, according to our theory of the variation of the barometer, there is
then much vapourized air: from all which it
feems probable, that air highly vapourized becomes an imperfect conductor, and, of courfe,
a difcharge made along a *ftratum* of it, will exhibit light, which I fuppofe to be the general cafe
of thunder and lightning.

Now, from the conclufions in the preceding
fections, we are under the neceffity of confidering
the *beams* of the *aurora borealis* of a *ferruginous*
nature,

nature, becaufe nothing elfe is known to be magnetic, and confequently, that there exifts in the higher regions of the atmofphere an elaftic fluid partaking of the properties of *iron,* or ra-ther of *magnetic fteel,* and that this fluid, doubt-lefs from its magnetic property, affumes the form of cylindric beams.—It fhould feem too, that the rainbow-like arches are a fort of *rings* of the fame fluid, which encompafs the earth's northern magnetic pole, like as the parallels of latitude do the other poles ; and that the beams are ar-ranged in equidiftant rows round the fame pole. At firft view, indeed, it feems incompatible with the known laws of magnetifm, that a quantity of magnetic matter fhould affume the form of fuch rings, by virtue of its magnetifm ; but it may take place in one cafe at leaft, if we fuppofe the rings fituate in the middle, between two rows of beams, fo that the attraction on each fide may be equal. As for the beams, in their natural ftate, when not acted upon by caufes hereafter to be mentioned, they muft all be guided by the *earth's magnetifm* (I mean the caufe that guides the needle, whether it is in the earth or air I know not), and confequently all have their *north poles* downward ; but whether any two neighbouring beams have the poles of the fame denomination, or of different denomi-nations, acting upon each other, ftill the effect will be the fame, and their mutual action upon each other not difturb their parallelifm, nor the

pofition

pofition of the rings; becaufe, whether the poles mutually attract or repel each other, is of no moment in this cafe, and the attraction of each pole is alike upon the rings.

Things being thus ftated, I moreover fuppofe, that this elaftic fluid of magnetic matter is, like vapourized air, an *imperfect conductor* of electricity; and that when the equilibrium of electricity in the higher regions of the atmofphere is difturbed, I conceive that it takes thefe beams and rings as conductors, and runs along from one quarter of the heavens to another, exhibiting all the phenomena of the *aurora borealis.*—The reafon why the diffufe flafhes fucceed the more intenfe light of the beams is, I conceive, becaufe the electricity difperfes the beams in fome degree, which collect again after the electric circulation ceafes.

Many of my readers, I make no doubt, will be furprifed to find, after having formed a conception that the relation betwixt the *aurora* and magnetifm was to be explained and demonftrated, chiefly if not folely, from the obfervations on the difturbance of the needle during the *aurora*, that no mention or ufe whatever is made of thofe obfervations, in the preceding fections. In fact, the relation above mentioned is demonftrable without any reference to them; notwithftanding which, they not only corrobo-

rate

rate the proof of it, but almoft eftablifh the truth of the hypothefis we are here advancing.

The variations of the needle during the *aurora,* as may be feen in the obfervations, are fo exceedingly irregular, that after confidering them a while, one would conclude this is the only fact afcertained by thefe obfervations. However, I think we may deduce the following:

1. When the *aurora* appears to rife only about 5, 10, or 15° above the horizon, the difturbance of the needle is very little, and often infenfible.

2. When it rifes up to the zenith, and paffes it, there never fails to be a confiderable difturbance.

3. This difturbance confifts in an irregular ofcillation of the horizontal needle, fometimes to the eaftward, and then to the weftward of the mean daily pofition, in fuch fort that the greateft excurfions on each fide are nearly equal, and amount to about half a degree each, in this place.

4. When the *aurora* ceafes, or foon after, the needle returns to its former ftation.

Now, from thefe facts alone, independent of what is contained in the preceding fections, I
think

think we cannot avoid inferring, that there is
fomething magnetic *conftantly* in the higher regions of the atmofphere, that has a fhare at leaft
in guiding the needle; and that the fluctuations
of the needle during the *aurora* are occafioned
by fome mutations that then take place in this
magnetic matter in the incumbent atmofphere;
for, it is certainly improbable, if not abfurd, to
fuppofe that the *aurora produces* this magnetic
matter, at its commencement, and *deftroys* it
at its termination. Moreover, abftracting from
a chymical folution of the metal, nothing is
known to affect the magnetifm of *fteel*, but *heat*
and *electricity;* heat weakens, or deftroys it;
electricity does more, it fometimes changes the
pole of one denomination to that of another, or
inverts the magnetifm. Hence, we are obliged
to have recourfe to one of thefe two agents, in
accounting for the mutatious above mentioned.
As for heat, we fhould find it difficult, I believe,
to affign a reafon for fuch fudden and irregular
productions of it in the higher regions of the
atmofphere, without introducing electricity as an
agent in thofe productions; but rather than
make fuch a fuppofition, it would be more philofophical to fuppofe electricity to produce the
effect on the magnetic matter *immediately.* Hence
then were we obliged to form an *hypothefis* of
the *aurora borealis,* without any other facts relative to it than the *four* above mentioned, we
ought to fuppofe it a phenomenon produced in
some

manner by the united agency of magnetifm and electricity.

It appears then, that the difturbance of the needle during an *aurora* equally countenances the conclufions drawn in the laft fection, and the hypothefis adopted in this; and it may be accounted for on the hypothefis, as follows.

The beams of the *aurora*, being magnetic, will have their magnetifm weakened, deftroyed, or inverted, *pro tempore*, by the feveral electric fhocks they receive during an *aurora*; or perhaps the temporary difperfion and diffufion of the magnetic matter thereby, may confiderably alter its influence; when, therefore, the alterations on each fide of the magnetic meridian do not balance each other, the confequence will be a difturbance of the needle*.

B b   In

* I conceive that a beam may have its magnetifm inverted, and exift fo for a time, becaufe the repulfion, acting longitudinally upon it, will only impel it in that direction, and not turn it round; juft as the north pole of a magnet may be applied to the north pole of a magnetic needle, without turning it round, by keeping the magnet exactly in the fame line with the needle, and thus making the needle act upon the centre. And I further conceive, that when the beam is reftored to its natural pofition of the north pole downward, it is effected, not by inverting the beam, wholly as a beam, (for this is never obferved in an *aurora*) but by inverting the conftituent particles, which may eafily be admitted of a fluid.

In fine, the conclusions in the last section, and the hypothesis in this, afford a very plausible reason for the appearance of the *aurora* being so much more frequent now than formerly in these parts; if the earth's magnetic poles be like the centres of the *aurora*, as the phenomena indicate, it is plain the *aurora* must move along with them, and appear or disappear at places, according as the magnetic poles approach or recede from them; and hence it may be presumed that the earth's magnetic pole in the northern hemisphere is nearer the west of *Europe* in this century than it was in the last or preceding.—The observations upon the dip of the needle, however, if they have been accurately made, seem to indicate the approach of the magnetic pole to have been very little; the dip at *London*, according to Mr. *Cavallo*, was 71° 50′ in 1576, and 72° 3′ in 1775; but there is reason to suspect the accuracy of the instruments at so early a period as 1576; besides, we do not know in what proportion the dip of the needle keeps pace with the approach of the pole.

It may perhaps be necessary here, before the subject is dismissed, to caution my readers not to form an idea that the *elastic fluid of magnetic matter*, which I have all along conceived to exist in the higher regions of the atmosphere, is the same thing as the *magnetic fluid or effluvia* of most writers on the subject of magnetism.

netifm. This laft they confider as the efficient caufe of all the magnetic phenomena; but it is a mere hypothefis, and the exiftence of the *effluvia* has never been proved My *fluid of magnetic matter* is, like magnetic fteel, a fubftance poffeffed of the properties of magnetifm, or, if thefe writers pleafe, a fubftance capable of being acted upon by the magnetic *effluvia,* and not the magnetic *effluvia* themfelves.

Whether any of the various kinds of air, or elaftic vapour, we are acquainted with, is magnetic, I know not, but hope philofophers will avail themfelves of thefe hints to make a trial of them.

—————

SECTION FIFTH.

An inveftigation of the fuppofed effect of the Moon in producing the Aurora Borealis*.

SOME time after the author began his ob- fervations on the *aurora borealis,* it occurred to him that the phenomenon had more frequently happened about the change of the moon than at

B b 2   any

* An effay on this fubject was firft publifhed by the au- thor in the beginning of 1789, in Mr. *Davifon's Mathematical and Philofophical Repofitory.*

any other time; this produced the fuspicion that the ærial tides occafioned by the moon might have fome influence upon it. Granting this to be the cafe, it was obvious, the full moon muft have an equal fhare with the new, though the phenomenon may often be then invifible, owing to the light of the moon ——Having now an enlarged lift of obfervations, we fhall refume the fubject afrefh, and examine what countenance the obfervations give to the fuppofition.

In the lift of obfervations we have placed the moon's age, both with refpect to change and full; collating, therefore, the whole number of obfervations to each particular number exprefing the age, we fhall have the following feries:

Days paft change and full.	0	1	2	3	4	5	6	7
No. of obfervations.	14	25	21	20	19	20	15	21

Days paft change and full.	8	9	10	11	12	13	14.
No. of obfervations.	18	23	15	6	10	13	9.

$$(12)$$

As the lunar revolution is completed in $29\frac{1}{2}$ days nearly, one half of a lunation is $14\frac{1}{4}$ days; hence the obfervations under 14 do not ftand the fame chance as the reft, there being only $\frac{1}{4}$ of the number of periods that have a day correfponding to this number: the number of obfervations under it ought therefore to be increafed in the ratio

ratio of ¼ to 1, or be 12 inſtead of 9, in order
to make a fair diviſion of the terms of the ſeries.
Now the ſpring tides will fall almoſt wholly in
the firſt half of this period, and the neap tides
in the laſt ; dividing the terms of the ſeries,
therefore, into two equal portions, taking half
of the odd intermediate one to each, the ſums
of the portions are as under.

	Spring tides.	Neap tides.
No. of *auroræ.*	144½.	107½.
Ratio	4 :	3, nearly, which

is favourable to the ſuppoſition.

It may be objected, that as the latter diviſion
contains the whole of the *ſecond* quarter of the
moon, when its light is ſtrong, and when it is
above the horizon all the time there is to obſerve
the *aurora,* the phenomenon is not noticed as
often as it takes place in that quarter.—This
may be right, but it ſhould be obſerved, that the
laſt quarter of the moon, which is wholly exempt
from this objection, falls in the ſame diviſion ;
and both the firſt and third quarters, conſtituting
the other diviſion, are in part liable to the ſame
objection.

However, in order to determine whether this
objection is of ſuch import as to counterbalance
the apparent concluſion contained above, it may
be proper to find and compare the number of
obſervations

obfervations in the firſt and laſt quarters only.—
This being done, on the principle above uſed,
the numbers ſtand,

Firſt quarter, or ſpring tides.　　Laſt quarter, or neap tides.

93½.　　　　　　　　81.

From which it appears the phenomenon is obſerved more frequently in the firſt quarter of the
moon, though liable in part to the above objection, than in the laſt quarter, which is wholly
free from it.

Preſuming then from what is done above, that
thoſe periods of the lunar months, when the
higher tides are in the air, are moſt ſubject to
the phenomenon in queſtion, it ſhould be expected, that thoſe times of the *day* when ſuch
tides are in the atmoſphere, ſhould likewiſe be
more ſubject to it than others. Now the ſpring
tides in the afternoon always happen in the interval from 2 to 8 o'clock; conſequently, the
opportunity of making obſervations upon the
phenomenon in this interval will *often* occur in
in winter, and *never* in ſummer, owing to the
twilight.—It ſhould ſeem then, that the winter
obſervations ought to favour the hypotheſis more
than the ſummer ones.—In fact, we find this the
caſe. The obſervations in the months of November, December, and January, being arranged
and ſummed up as above, give,

Spring

Spring.	*Neap.*
40½.	24½.

And thofe in the months of May, June, July, and Auguft, give,

Spring.	*Neap.*
25½.	24½.

As the tides are higher in fpring and autumn than in fummer and winter, the phenomenon ought, according to hypothefis, to occur more frequently in the two former feafons than in the two latter. The number of obfervations in the different months ftand thus :

Jan.	Feb.	Mar.	April	May	June	July	Aug.	Sep.	Oct.	Nov.	Dec.
18	18	26	32	21	5	2	21	23	36	38	9.

The fmall number in June and July is undoubtedly owing in great part to the twilight ; but the deficiency in December, January, and February, cannot be owing to the fame caufe.

Upon the whole, I think it is not improbable that the agitations caufed by the moon in the very high regions of the atmofphere, which we may fuppofe are not much agitated by the tempefts in the lower regions, may have fome effect upon the phenomenon in queftion ; and the fuppofition is evidently countenanced by the feveral facts ftated above.

SECTION

SECTION SIXTH.

An inveſtigation of the effeƈt of the Aurora Borealis *on the Weather ſucceeding it.*

VARIOUS have been the conjeƈtures on this ſubjeƈt offered to the conſideration of the public: ſome aſſert that the *aurora* has no ſenſible effeƈt upon the weather; others that it is very frequently followed by rain ſoon after.

In the American Philoſophical Tranſaƈtions, we find it obſerved that the barometer *falls* after an *aurora.*

Having a large number of obſervations on the *aurora,* together with thoſe on the barometer and rain, we are prepared to examine theſe opinions, and we do it the rather becauſe if any thing can be aſcertained on this head, it muſt be regarded as a valuable diſcovery, conſidering the preſent very imperfeƈt ſtate of meteorological prognoſtication.

Since the ſpring of 1787 there have been 227 *auroræ* obſerved at *Kendal* and *Keſwick;* 88 of the next ſucceeding days were *wet,* and 139 *fair,* at *Kendal;* now, in the account of rain, the mean yearly number of wet days there is ſtated at 217, and of courſe the fair days are 148; hence the chances of any one day, taken at random,

dom, being wet or fair, are as thefe numbers. But it appears the proportion of fair days to wet ones fucceeding the *aurorae*, is much greater than this general ratio of fair days to wet ones; the inference therefore is, that the appearance of the *aurora borealis* is a prognoftication of *fair weather.*

The only objection to this inference which occurs to us as worth notice is, that the *aurora* being from its nature only vifible in a clear atmofphere, this circumftance of itfelf is fufficient to caft the fcale in favour of the fucceeding day being fair, without confidering the *aurora* as having any influence either directly or indirectly. —This objection has undoubtedly fome weight; but upon examinining the obfervations, it appears that the *aurora* not only favours the next day, but indicates that a feries of days to the number of 10 or 12, are more likely to be all fair, than they would be without this circumftance.

Of 227 obfervations, 139 were followed by 1 or more fair days, 100 by 2 or more &c. as under.

1	2	3	4	5	6	7	8	9	10	11	12
139	100	69	52	38	30	21	16	10	6	2	1.

According to the laws of chance, the probability of any number of fucceffive fair days is found by raifing $\frac{148}{365}$ to the power, whofe index is the propofed number of fair days; thefe probabilities being multiplied by 227 will give what

C c   the

the above feries ought to have been, if the *aurora* had no influence; it is as under.

1	2	3	4	5	6
92	38	15	6	2	1

From which it appears, there fhould not have
been more than 1 *aurora* out of 227 followed
by 6 fair days, and yet in fact there were 39;
whence the inference above made is confirmed.

As for the different feafons of the year, I find
the *aurora* is more frequently followed by fair
weather in fummer than in winter; but the dif-
tinction is not very confiderable.

It may be obferved that the largeft and moft
fplendid appearances of the *aurora*, as they ufually happen in rainy and unfettled weather, they
are frequently fucceeded by 1 or more wet days;
but I do not find any of thofe very remarkable
ones which happened on a fair day, was fucceeded by a wet one.

Upon examination of the effect of the *aurora*
upon the barometer, I find, that fince the 19th
of September, 1787, there have been 219 obfervations, and that in 120 of thefe inftances the
barometer was rifen next morning, and fallen in
99.—This circumftance, therefore, corroborates
the inference before made, that the *aurora* is a
fign of fair weather.

General

General Rules and Observations for judging of the Weather.

NOTWITHSTANDING we have departed pretty much from our original defign of expatiating on this fubject, we think it may not be amifs to collect fome of the facts and obfervations that are diffufed through the work, which relate more immediately to the fubject, and to add thereto a few more obfervations.

1. The barometer is higheft of all during a long froft, and generally rifes with a NE. wind ; it is loweft of all during a thaw following a long froft, and is often brought down by a SW. wind. See page 112.

2. When the barometer is near the high extreme for the feafon of the year, there is very little probability of immediate rain. See page 151.

3. When the barometer is low for the feafon, there is feldom a great weight of rain, though a fair day in fuch a cafe is rare. See page 150. The general tenor of the weather at fuch times is, fhort, heavy, and fudden fhowers, with fqualls of wind from the SW. W. or NW.

4. In fummer, after a long continuance of fair weather, with the barometer high, it generally falls gradually, and for one, two, or more days, before there is much appearance of rain.—If the

fall

fall be fudden and great for the feafon, it will probably be followed by thunder.

5. When the appearances of the fky are very promifing for fair, and the barometer at the fame time low, it may be depended upon the appear. ances will not continue fo long. The face of the fky changes very fuddenly on fuch occafions.

6. Very dark and denfe clouds pafs over with. out rain when the barometer is high; whereas, when the barometer is low, it fometimes rains almoft without any appearance of clouds.

7. All appearances being the fame, the higher the barometer is, the greater the probability of fair weather.

8. Thunder is almoft always preceded by hot weather, and followed by cold and fhowery weather.

9. A fudden and extreme change of tempera. ture of the atmofphere, either from heat to cold or cold to heat, is generally followed by rain within 24 hours.

10. In winter, during a froft, if it begin to fnow, the temperature of the air generally rifes to 32°, and continues there whilft the fnow falls; after which, if the weather clear up, expect fe- vere cold.

11. The *aurora borealis* is a prognoftic of fair weather. See Effay 8, Sect. 6.

Appendix,

(197)

Appendix, containing additional Notes, &c. on different parts of the Work.

PAGE 8.

THE height of *Kendal* above the fea was fet down 25 yards, by eftimation only; I have fince found, by levelling with the barometer, that *Stramongate bridge*, at *Kendal*, is 46 yards above *Levens bridge*, to which the tide flows; though it feems the furvey for the intended canal makes the height lefs: I am not aware of any circumftance that could lead me into an error.

Mr. *Crofthwaite* has lately determined, by levelling with a very good theodolite, that *Baffenthwaite-lake* is 70 yards above the level of the fea, and that *Derwent-lake*, which is 10 yards below his barometer, is 76 yards above the level of the fea; I make the laft mentioned lake 81 yards above the level of the fea, from barometrical obfervations; but if I have made an error by determining *Kendal* 5 yards too high, the refults of our obfervations will be reconciled *.

Page

* The height of the following places above the level of the fea have been determined as under; the obfervations with the barometer were made by myfelf, and thofe with the theodolite by Mr. *Crofthwaite*.

	From barom. obf.	From the theodolite.
Windermere lake - - - -	26 yards.	————
Dunmail-raife, barrow of ftones in the boundary of *Cumberland* and *Weftmorland*	245 ——	275.
Leathes lake - - - - -	171 ——	182.

My

Page 29.—The greateſt heat experienced for the laſt 5 years, at *Kendal*, was on the 1ſt of Auguſt 1792; but the heat of the preſent year, 1793, exceeded; the thermometer in the ſhade was 83°$\frac{1}{2}$ on the 11th, and 84$\frac{1}{2}$ on the 15th of July.

Page 39.—There is a great diſcordance in the height of *Skiddaw*, as determined by the obſervations of different perſons; I have remarked that Mr. *Croſthwaite* made it 1050 yards above *Derwent lake*, I find ſince that Mr. *Donald* made it 1090 yards above the ſea, and 958 above *Baſſenthwaite lake*. Mr. *Croſthwaite*, by a later admeaſurement, determines its height 1000 yards above *Derwent lake*.—On the 26th of Auguſt, 1793, I attempted its height by the barometer, for which purpoſe the following obſervations were made.

At 3h 30m P.M. the barometer upon the ſummit of *Skiddaw*, when the proper allowance for the riſe of the mercury in the reſervoir was made, ſtood at - - - - - - - - - 26.79 inches.
Mr. *Croſthwaite*'s barometer at *Keſwick*, allowing for the ſmall difference of the barometers when together, at the ſame time, ſtood at - - 29.715——.
A detached thermometer above was, in the ſhade, - 46°.
A detached thermometer below was, in the ſhade, - 60°.

Now,

My obſervations were taken both in going to and returning from *Keſwick*, and compared with nearly cotemporary obſervations at *Kendal* and *Keſwick*; at the former time the air was dry, and at the latter moiſt: the elevations were found ſomething leſs by the later obſervations, but the difference was only 3 yards in *Leathes lake* and 9 in *Dunmail-raiſe*.

Now, by applying the theorem at page 81, we find the elevation of the upper barometer above the lower 945½ yards; whence, adding 10¾ yards, we get the height of *Skiddaw* above *Derwent lake* = 956¼ yards, and its height above the fea comes out 1037¼ yards.—It had been a good deal of rain on the morning of that day, and the clouds were juft broken off at the time of the obfervations, the air remaining ftill very foft; from which circumftance I am inclined to think that the height above determined is rather too little; for I have found by repeated obfervations upon a hill 310 yards high, that the heights are found lefs by the theorem as the air is fofter, *cæteris paribus:* I think therefore we may conclude *Skiddaw* to be nearly 1000 yards above *Derwent lake*, agreeable to Mr. *Crofthwaite's* laft meafurment, till its height can be more exactly afcertained by a repetition of obfervations *.

———

AFTER

* Mr. *Crofthwaite* makes the height of *Latrig*, another mountain in the neighbourhood of *Kefwick*, to to be 319 yards above *Derwent lake:* by obfervations on the barometer the above mentioned day, I found its height 312 yards, which, for the reafon affigned above, is probably too little.

Helvellyn is a mountain clofe by the road leading from *Kendal* to *Kefwick*, about 8 miles from the latter place; it has always juftly been confidered higher than *Skiddaw*. On the 27th of Auguft I made the following obfervations to determine its height.

At 1h 30m P. M. barometer at the fummit, corrected as above, 26 69.
Barometer below, 10 yards above *Leathes lake*, - - - - - 29.39.
A detached thermometer at the fummit was - - - - - - 42°½.
A detached thermometer below was - - - - - - - 54.

From

AFTER I had obferved the *aurora borealis* to to difturb the needle fo greatly, as is related in the *addenda* to the obfervations on that head, I conjeftured, *a priori,* that thunder-ftorms would do the fame; accordingly, I watched the needle for a confiderable time during the only thunder-ftorm we had at *Kendal* in the fummer of 1793, namely, on the evening of the 3d of Auguft; but, far from perceiving any unufual fluftuation, I could not difcover the needle was perceptibly difturbed all the while, and it continued at the fame ftation the next morning.

On the ftate of Vapour in the Atmofphere, &c.—
See page 134, *and following.*

AFTER making fome experiments upon the effefts of the condenfation of atmofpheric air, in a glafs veffel, by means of a fyringe, from which I find that repeated condenfation produces a de-pofition of water upon the infide of the glafs, and repeated rarefaftion removes the fame; alfo,
having

From which the elevation of the upper barometer above the lower comes out $869\frac{1}{3}$ yards; to which adding 171 and 10, we get the height of *Helvellyn* above the fea $= 1050\frac{1}{3}$ yards. But it fhould be obferved the ftate of the air was ftill more moift than when I was upon *Skiddaw,* and the obfervation at top was taken during a fhower; from which it is probable the height of *Helvellyn* above the fea is nearly 1100 yards: Mr. *Donald* makes it 1108 above the fea.—About 200 yards below the fummit there is a very fine fpring, from which a large ftream of water defcends all the year round, with little variation in quantity at the different feafons, as my guide in-formed me; its temperature I found to be 38°.

having made fome experiments upon the effect
of heat on water thrown into the *vacuum* of a
common barometer, which tend to confirm thofe
the refult of which is given at page 134,—I am
confirmed in the opinion, that *the vapour of water (and probably of moft other liquids*) exifts at
all times in the atmofphere, and is capable of bearing
any known degree of cold without a total condenfation, and that the vapour fo exifting is one and the
fame thing with* fteam, *or vapour of the temperature
of* 212° *or upwards.* The idea, therefore, that
vapour cannot exift in the open atmofphere
under the temperature of 212°, unlefs chymically combined therewith, I confider as errone-
ous; it has taken its rife from a fuppofition that
air preffing upon *vapour* condenfes the vapour
equally with *vapour* preffing upon *vapour*, a fuppofition we have no right to affume, and which
I apprehend will plainly appear to be contradictory to reafon, and unwarranted by facts; for,
when a particle of vapour exifts between two
particles of air, let their equal and oppofite pref-
fures upon it be what they may, they cannot
bring it nearer to another particle of vapour,
without which no condenfation can take place,
all other circumftances being the fame; and it has
never been proved that the vapour in a receiver
from which the air has been exhaufted is precipitated upon the admiffion of perfectly dry air.
Hence, then, we ought to conclude, till the con-

D d   trary

---

* Dr. *Prieftley* obferves in the fifth volume of his Experiments, page 225, that quickfilver evaporates not only *in vacue*
but when expofed to the atmofphere.

trary can be proved, that *the condenfation of vapour expofed to the common air, does not in any
manner depend upon the preffure of the air.*

All the facts, however, confpire to prove that
the *temperature* of the air bears a relation to the
condenfation of vapour; thus, the utmoft force
which vapour of 212° can exert, is equivalent
to the weight of 30 inches of mercury, and any
greater force than this, acting upon vapour alone
of that temperature, will condenfe the whole into
water; and, if the temperature be lefs, then the
utmoft force or fpring of vapour is lefs, as is indicated by the table in page 134; and no doubt
as the utmoft force decreafes, the utmoft denfity
will decreafe alfo, though probably not in the fame
ratio.   Hence, then, atmofpheric air, faturated
with vapour, is fuch wherein the vapour, confidered abftractedly from the air in which it is
diffufed, is at its utmoft denfity for the temperature; in fuch cafe, if a quantity of atmofpheric
air and vapour be taken, and mechanically condenfed, a portion of the vapour will be condenfed
into water, and give off heat; on the contrary,
if it be expanded, or, which amounts to the
fame thing, if a quantity be taken out of a receiver, the remainder will have its capacity for
vapour increafed, as has been already obferved.

Though the preffure of the air does not promote the condenfation of vapour, yet when the
preffure is removed, evaporation is promoted;
for under the receiver of an air-pump we find
that

that the vapour from the wet leather rifes as faft as it can be pumped out, when the rarefaction has proceeded to a certain degree.

In order the more to illuftrate and confirm the notion of vapour here laid down, we fhall now attempt to explain feveral facts, which have been confidered as involving difficulties, and we believe fome of them have never been accounted for by others.

Dr. *Alexander*, in his *Experimental Effays*, page 102, informs us, that from fome experiments he was induced to think, that blowing upon the bulb of a thermometer with a pair of hand-bellows would cool it, but upon trial found it was always heated 1 or more degrees by the operation —Now, if a thermometer that has juft been dipped in water of the fame temperature as the air, be blown with a pair of hand-bellows as above, it will be cooled feveral degrees. Thefe two facts I have proved frequently, from expement.—Again, Dr. *Darwin* (fee the note, page 136) found that air having been for fome time condenfed, upon rufhing out againft the bulb of a thermometer, cooled it feveral degrees, and a dew was depofited upon the bulb.

The reafon of thefe apparently difcordant facts may be explained thus: the condenfation of vapour in a pair of hand-bellows will precipitate a portion of the infufed vapour, which gives off its heat to the air; and thus the temperature of the

D d 2

air

air in the bellows being increafed, that of the
thermometer, expofed to the current, will be
increafed accordingly. In the fecond inftance,
the water on the bulb of the thermometer being
expofed to the current of air, quickly evaporates, and at the fame time abforbs the neceffary
heat from the quickfilver. But in the third inftance the heat confequent to the condenfation
was fuffered to efcape, whilft the condenfed vapour or water remained in the air-gun; the air
rufhing out was therefore of the fame temperature as the furrounding air, and probably a great
portion of the condenfed vapour remained mechanically mixed therewith ; a depofition of water upon the bulb of the thermometer was of
courfe unavoidable, and this being refolved into
vapour by its expofition, reduced the temperature
of the thermometer.

In the Philofophical Tranfactions for 1777,
there is a very interefting feries of experiments
fhewing the effects of vapour in the receiver of
an air-pump, when the air is exhaufted ; the
experiments were made by *Edward Nairne,*
F. R. S. upon a pump on Mr. *Smeaton's* con-
ftruction. He ufed two gages, one of which
was the common barometer gage, which was of
courfe an accurate meafure of the force or elafticity of the medium of air or vapour within the
receiver ; the other, called the *pear gage,* from
its fhape, confifted of a glafs tube, capacious in
the middle, and ending in a narrow neck, which
was

was clofe; the other, or open end, was, by a contrivance for the purpofe, let into a bafon of mercury before the air from without was fuffered to enter, and upon its admiffion the quickfilver was forced into the gage; the fpace occupied by the air being then compared with the whole capacity of the gage, gave the rarefaction of the permanent elaftic fluid or air.—The chief facts obferved were the following,

1. When the pump-plate leather was foaked in water, and the barrel of the pump well cleared of moifture, then, after working the pump for 10 minutes, the rarefaction indicated by the pear gage was very great, and exceeded what was obferved in any other circumftance, whilft that indicated by the barometer gage was often not $\frac{1}{700}$th part as great as the other; alfo, it was obferved that the rarefaction by the pear gage was *lefs* every time the experiment was repeated, but that of the barometer gage was always the *fame* at the fame time.

2. When the pump-plate leather was foaked in water mixed with fpirit of wine, the rarefaction by both gages was lefs than in the former cafe; but the refults in other refpects were fimilar.

3. The effects of different temperatures of the air upon the rarefaction were as follow :

Pump-plate leather being foaked in water.

Air in the room 46°—barometer gage 84—pear gage 20000.
——————— 57 ——————56 ———— 16000.

Pump-plate

Pump-plate leather being soaked in water mixed with spirit of wine,
Air in the room 46°—barometer gage 76—pear gage 8000.
———— 57 ———— 49 ———— 1200.

4. When the leathers of the *piston* were soaked in water, the two gages nearly corresponded; but the utmost rarefaction in this circumstance was very small, being, for instance,

In one pump — barometer gage 37——pear gage 38.
In another pump ———— 34 ———— 37.

5. When the pump, &c. were dry, the barometer gage was sometimes lower after working the pump 5 minutes, than after the operation was continued 5 minutes longer.

6. When the pump and plate were both dry, and the receiver cemented on to the pump-plate, the two gages nearly agreed, the rarefaction by both being about 600, in *damp* weather; but in *dry* weather, and in a still greater degree when a quantity of vitriolic acid was in the receiver, (which was always found to gain weight by such its exposure) the barometer gage indicated a greater rarefaction than the pear gage.

These facts, some of which the ingenious artist who made the experiments accounted for, seem most or all of them capable of a satisfactory explanation upon the theory of vapour we are here maintaining.—When the pump-plate leather is soaked in any liquid, and the pressure is so far diminished that the liquid boils, or turns into vapour, it is plain the pressure can be no further
diminished;

diminifhed; and in fuch cafe, when the pump is wrought, it muft draw each time a portion of the remaining air along with the vapour, and thus the air in the receiver admits of a diminution almoft *ad infinitum*, and vapour generated inftantaneoufly fupplies the place of the air withdrawn; when air is let in, the vapour in the pear gage is condenfed, and there remains nothing but the fmall portion of air, with its faturating portion of vapour, at the top of the gage.—The reafon why the repetition of the experiment decreafed the rarefaction by the pear gage, was, that the frequent condenfations of air and vapour in the barrel of the pump muft have produced a depofition of water there, by which the effect was fooner at its *ne plus ultra;* for, when the vacuum of the barrel is not perfect, the quantity drawn from the receiver in a given time muft be lefs than otherwife. I have no doubt if the experiments had been repeated often enough, the leather of the pifton and the valves would have been in effect foaked with water, and the refult as ftated in the 4th fact: in this cafe, as foon as the fpring of the air in the receiver is weakened to a certain degree, working the pump does not avail, becaufe the vapour in the barrel, together with the refiftance of the valves, is juft fufficient to counteract the fpring of the air within; hence the rarefaction by the pear gage is then fcarcely greater than by the barometer gage.

Experience proves that fpirit of wine rifes fooner into vapour than water; confequently the
rarefaction

rarefaction by the pear gage, when the pump is wrought a given time, muſt be leſs than when water is uſed. Alſo, it follows *a priori*, that the cooler the circumambient air, other circum-ſtances being the ſame, the greater muſt the rarefaction be by both gages.

When by long pumping a quantity of vapour is collected in the barrel of the pump, I con-ceive a portion of it may, during the operation, eſcape again into the receiver, this will account for the 5th fact.

I do not ſee how the 6th fact can be explained without ſuppoſing that the elaſticity of dry air, when greatly expanded, decreaſes in a greater proportion than its denſity; it is true that the increaſe of cold in the receiver, and the leſs va-pour there is in the air at firſt, the more will the rarefaction indicated by the barometer gage ex-ceed that of the pear gage; for, it cannot be reaſonably ſuppoſed that when the rarefaction is at its utmoſt degree, the proportion of vapour to air in the receiver is no greater than at firſt; I conceive, therefore, that the air condenſed in the pear gage is always ſaturated with vapour, unleſs perhaps when the vitriolic acid is in the receiver, and of courſe its bulk, *cæteris paribus*, greater than before: but this alone is not ſufficient to account for the obſerved differences of the gages.

THE END.

Printed in the United States
By Bookmasters